S0-AOJ-735

VOLUME FIVE HUNDRED AND THIRTY NINE

METHODS IN ENZYMOLOGY

Laboratory Methods in Enzymology: Protein Part B

METHODS IN ENZYMOLOGY

Editors-in-Chief

JOHN N. ABELSON and MELVIN I. SIMON
Division of Biology
California Institute of Technology
Pasadena, California

ANNA MARIE PYLE
Departments of Molecular, Cellular and Developmental
Biology and Department of Chemistry
Investigator, Howard Hughes Medical Institute
Yale University

Founding Editors

SIDNEY P. COLOWICK and NATHAN O. KAPLAN

VOLUME FIVE HUNDRED AND THIRTY NINE

METHODS IN ENZYMOLOGY

Laboratory Methods in Enzymology: Protein Part B

Edited by

JON LORSCH
Johns Hopkins University School of Medicine
Baltimore, MD, USA

AMSTERDAM • BOSTON • HEIDELBERG • LONDON
NEW YORK • OXFORD • PARIS • SAN DIEGO
SAN FRANCISCO • SINGAPORE • SYDNEY • TOKYO
Academic Press is an imprint of Elsevier

Academic Press is an imprint of Elsevier
525 B Street, Suite 1800, San Diego, CA 92101-4495, USA
225 Wyman Street, Waltham, MA 02451, USA
Radarweg 29, PO Box 211, 1000 AE Amsterdam, The Netherlands
The Boulevard, Langford Lane, Kidlington, Oxford, OX5 1GB, UK
32 Jamestown Road, London NW1 7BY, UK

First edition 2014

ISBN: 978-0-12-420120-0
ISSN: 0076-6879

Printed and bound in United States of America
14 15 16 17 11 10 9 8 7 6 5 4 3 2 1

CONTENTS

CONTRIBUTORS

Alice Alegria-Schaffer
Thermo Fisher Scientific, Rockford, IL, USA

Manuel Ascano
Howard Hughes Medical Institute, Laboratory of RNA Molecular Biology, The Rockefeller University, New York, NY, USA

Cathleen M. Brdlik
Department of Genetics, Stanford University, Stanford, CA, USA

Lukas Burger
Biozentrum der Universität Basel and Swiss Institute of Bioinformatics (SIB), Basel, Switzerland

Nicholas K. Conrad
Department of Microbiology, University of Texas Southwestern Medical Center, Dallas, TX, USA

Thalia Farazi
Howard Hughes Medical Institute, Laboratory of RNA Molecular Biology, The Rockefeller University, New York, NY, USA

Rachel Green
Department of Molecular Biology and Genetics, Johns Hopkins University School of Medicine, Baltimore, MD, USA

Markus Hafner
Howard Hughes Medical Institute, Laboratory of RNA Molecular Biology, The Rockefeller University, New York, NY, USA

Mohsen Khorshid
Biozentrum der Universität Basel and Swiss Institute of Bioinformatics (SIB), Basel, Switzerland

Yvonne Y. Koh
Microbiology Doctoral Training Program, University of Wisconsin – Madison, Madison, WI, USA

Markus Landthaler
Berlin Institute for Medical Systems Biology, Max-Delbruck-Center for Molecular Medicine, Berlin, Germany

Lauren Makuch
Department of Neuroscience, Johns Hopkins University School of Medicine, Baltimore, MD, USA

Wei Niu
Department of Genetics, Stanford University, Stanford, CA, USA

Jeff Nusbaum
Howard Hughes Medical Institute, Laboratory of RNA Molecular Biology, The Rockefeller University, New York, NY, USA

Matthew S. Sachs
Department of Biology, Texas A&M University, College Station, TX, USA

Emi Sei
Department of Microbiology, University of Texas Southwestern Medical Center, Dallas, TX, USA

Michael Snyder
Department of Genetics, Stanford University, Stanford, CA, USA

Jessica Spitzer
Howard Hughes Medical Institute, Laboratory of RNA Molecular Biology, The Rockefeller University, New York, NY, USA

Sarah H. Stubbs
Department of Microbiology, University of Texas Southwestern Medical Center, Dallas, TX, USA

Thomas Tuschl
Howard Hughes Medical Institute, Laboratory of RNA Molecular Biology, The Rockefeller University, New York, NY, USA

Greg Wardle
Howard Hughes Medical Institute, Laboratory of RNA Molecular Biology, The Rockefeller University, New York, NY, USA

Marvin Wickens
Department of Biochemistry, University of Wisconsin – Madison, Madison, WI, USA

Cheng Wu
Department of Biology, Texas A&M University, College Station, TX, USA

Hani S. Zaher
Department of Molecular Biology and Genetics, Johns Hopkins University School of Medicine, Baltimore, MD, USA

Mihaela Zavolan
Biozentrum der Universität Basel and Swiss Institute of Bioinformatics (SIB), Basel, Switzerland

Methods in Enzymology volumes provide an indispensable tool for the researcher. Each volume is carefully written and edited by experts to contain state-of-the-art reviews and step-by-step protocols.

In this volume we have brought together a number of core protocols concentrating on Protein, complimenting the traditional content which is found in past, present and future Methods in Enzymology volumes.

PREFACE

These volumes of *Methods in Enzymology* contain the protocols that made up the on-line *Methods Navigator.* Our philosophy when we selected the protocols to include in the *Navigator* was that they should be for techniques useful in any biomedical laboratory, regardless of the system the lab studies. Each protocol was written by researchers who use the technique routinely, and in many cases by the people who actually developed the procedure in the first place. The protocols are very detailed and contain recipes for the necessary buffers and reagents, as well as flow-charts outlining the steps involved. Many of the chapters have accompanying videos demonstrating key parts of the procedures. The volumes are broken into distinct areas: DNA methods; Cell-based methods; lipid, carbohydrate and miscellaneous methods; RNA methods; Protein methods. Our goal is that these protocols will be useful for everyone in the lab, from undergraduates and rotation students to seasoned post-doctoral fellows. We hope that these volumes will become dog-eared and well-worn in your laboratory, either physically or electronically.

Professor Jon Lorsch
Johns Hopkins University
School of Medicine

SECTION I

Protein Protocols/ Protein *In Vitro* Translation

CHAPTER ONE

In Vitro Synthesis of Proteins in Bacterial Extracts

Hani S. Zaher, Rachel Green[1]
Department of Molecular Biology and Genetics, Johns Hopkins University School of Medicine, Baltimore, MD, USA
[1]Corresponding author: e-mail address: ragreen@jhmi.edu

Contents

Abstract

This protocol describes the methods used to generate protein in a cell-free system derived from *E. coli*. The *in vitro* synthesis of protein has been used in studying many ribosome-based gene regulation steps (Gong and Yanofsky, 2001). Such techniques have also been used to study protein–protein, protein–DNA, and protein–RNA interactions, and to produce radiolabeled protein species. More recently, such approaches

Methods in Enzymology, Volume 539
ISSN 0076-6879
http://dx.doi.org/10.1016/B978-0-12-420120-0.00001-3

have been utilized to produce large quantities of toxic proteins for proteomic and structural studies (Yokoyama et al., 2000).

1. THEORY

Cell-free protein synthesis systems have been heavily utilized when its *in vivo* counterparts may fail or prove to be difficult to use. The advantages of *in vitro* protein synthesis are threefold: the environment of protein production can be readily manipulated by the researchers for optimal conditions, cell viability is not a prerequisite so toxic proteins can be studied, and lastly, large amounts of protein can be produced. Nonetheless, because this is an *in vitro* system (and may represent an artificial condition), it should be used only if carrying out the experiments in living cells is not feasible. Moreover, because of the complexity of the system, it should be a last resort for protein overexpression.

2. EQUIPMENT

French press
Shaking incubator
UV/vis spectrophotometer
High capacity centrifuge
Refrigerated high speed centrifuge
Microcentrifuge
Scintillation counter
Water bath
Polyacrylamide gel electrophoresis equipment
Film developer or Phosphorimager
Side-arm filter flask
Stainless steel filter holder
Erlenmeyer flasks, 250 ml and 4 l
Micropipettors
Micropipettor tips
Pipette aid
Polycarbonate centrifuge tubes, 50 ml
50-ml polypropylene tubes
1.5-ml microcentrifuge tubes
Whatman GF/C glass fiber filters
Autoradiography film or phosphorimager cassette

3. MATERIALS

Acetone
Potassium hydroxide (KOH)
Sodium hydroxide (NaOH)
Bacto tryptone
Bacto yeast extract
Potassium acetate (KOAc)
Potassium glutamate
Ammonium acetate (NH_4OAc)
Magnesium acetate (MgOAc)
[^{35} S]-Methionine (or other radiolabeled amino acid)
Tris base
Glacial acetic acid
HEPES
Sodium dodecyl sulfate (SDS)
Trichloroacetic acid (TCA)
Sodium chloride (NaCl)
EDTA
Phenylmethylsulfonyl fluoride (PMSF)
T7 RNA polymerase
Creatine kinase (CK)
Creatine phosphate (CP)
Pyruvate kinase (PK)
Phosphoenol pyruvate (PEP)
E. coli total tRNA mix
Bovine serum albumin
ATP
GTP
CTP
UTP
Amino acids
L(-)-5-Formyl-5,6,7,8-tetrahydrofolic acid (folinic acid)
cAMP
Dithiothreitol (DTT)
β-Mercaptoethanol
Polyethylene glycol 8000 (PEG 8000)
40% acrylamide/bisacrylamide (19:1)

3.1. Solutions & buffers

Step 1 2× YT media

Component	Amount
Bacto Tryptone	16 g
Bacto Yeast Extract	10 g
NaCl	5 g

Add water to 1 l. Adjust pH to 7.0 with 5 N NaOH, and autoclave

Buffer A

Component	Final concentration	Stock	Amount
Tris–acetate, pH 8.2	10 mM	1 M	10 ml
Magnesium acetate	14 mM	1 M	14 ml
Potassium acetate	60 mM	3 M	20 ml
DTT	1 mM	1 M	1 ml

Add water to 1 l

Step 2 Activation buffer

Component	Final concentration	Stock	Amount
Tris–acetate, pH 8.2	293 mM	1 M	5.9 ml
Magnesium acetate	9.2 mM	1 M	0.18 ml
ATP	13.2 mM	0.5 M	0.53 ml
PEP	84 mM	0.5 M	3.4 ml
DTT	4.4 mM	1 M	88 μl
Amino acids	40 μM each	10 mM each	80 μl
PK	6.7 U ml^{-1}	2000 U ml^{-1}	67 μl

Add water to 20 ml

Step 3 S30 cocktail

Component	Final concentration	Stock	Amount
HEPES–KOH, pH 7.5	184 mM	1 M	184 μl
DTT	5.7 mM	1 M	5.7 μl

ATP	4 mM	0.5 M	8 μl
CTP	2.7 mM	0.5 M	5.4 μl
GTP	2.7 mM	0.5 M	5.4 μl
UTP	2.7 mM	0.5 M	5.4 μl
CP	267 mM	1 M	267 μl
CK	0.8 mg ml^{-1}	20 mg ml^{-1}	40 μl
PEG 8000	13%		130 mg
Cyclic-AMP	2 mM	40 mM	50 μl
Folinic acid	0.2 mM	10 mM	20 μl
Total tRNA mixture	580 μg ml^{-1}	10 mg ml^{-1}	58 μl
Potassium glutamate	700 mM	3 M	233 μl
Ammonium acetate	90 mM	3 M	30 μl
Magnesium acetate	Variable	1 M	variable

Add water to 1 ml

Amino acids mixture
Make a 10 mM stock of the 19 amino acids (minus the labeled amino acid). Adjust the pH of the mixture to 7.5 using 1 M KOH

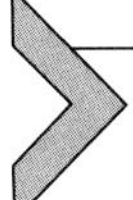

4. PROTOCOL

4.1. Duration

Preparation	**About 1 day**
Protocol	About 2 days

4.2. Preparation

Inoculate a single colony of *E. coli* into 50 ml of 2× YT medium and grow at 37 °C overnight with shaking (250 rpm). We typically use the BL21 CP strain.

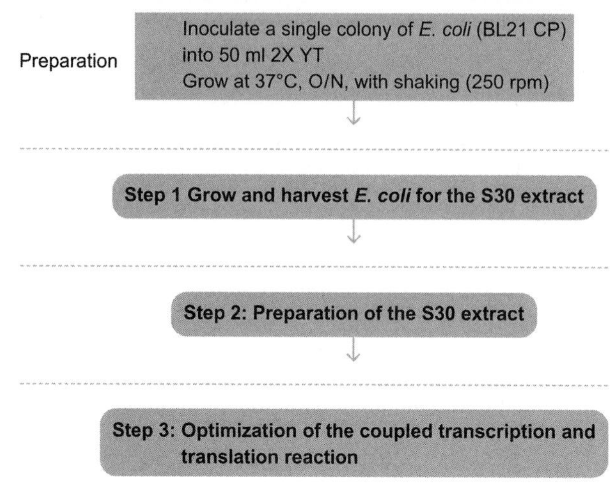

Figure 1.1 Flowchart of complete protocol, including preparation.

4.3. Caution

Consult your institute Radiation Safety Officer for proper ordering, handling, and disposal of radioactive materials.

See Fig. 1.1 for the flowchart of the complete protocol.

5. STEP 1 GROW AND HARVEST *E. COLI* FOR THE S30 EXTRACT

5.1. Overview

Grow *E. coli* until the culture reaches mid-logarithmic phase and harvest the cells for the preparation of the extract.

5.2. Duration

3–4 h

1.1 Start three cultures of 1 l each in 4-l Erlenmeyer flasks by inoculating 10 ml of the overnight culture into each flask.

1.2 Grow the culture at 37 °C until they reach mid-log phase (OD_{600} = 0.9–1.2).

1.3 Quickly chill the cells by immersing the flasks in an ice-water bath.

1.4 Centrifuge the cells, preferably using high capacity rotors (ones that can hold 0.5–1 l of liquid), at 6000 × g, 4 °C for 15 min.

1.5 Decant the spent media and, using a Pipet-aid, carefully resuspend each cell pellet originating from 1 l of culture in 20 ml of ice-cold Buffer A supplemented with 6 mM β-mercaptoethanol.

1.6 Transfer the resuspended cells into prechilled 50-ml centrifuge tubes before centrifuging at 6000 × g, 4 °C for 15 min.

1.7 Wash the cell pellet one more time with ice-cold Buffer A and centrifuge as before.

1.8 At this stage, the cell pellet can be quickly frozen by immersing it in liquid nitrogen and stored at −80 °C.

5.3. Tip

The freezing step can affect the overall activity of the extract and, as a result, it is preferable to carry on the next step immediately. Furthermore, the cells should never be frozen for more than 3 days.

See Fig. 1.2 for the flowchart of Step 1.

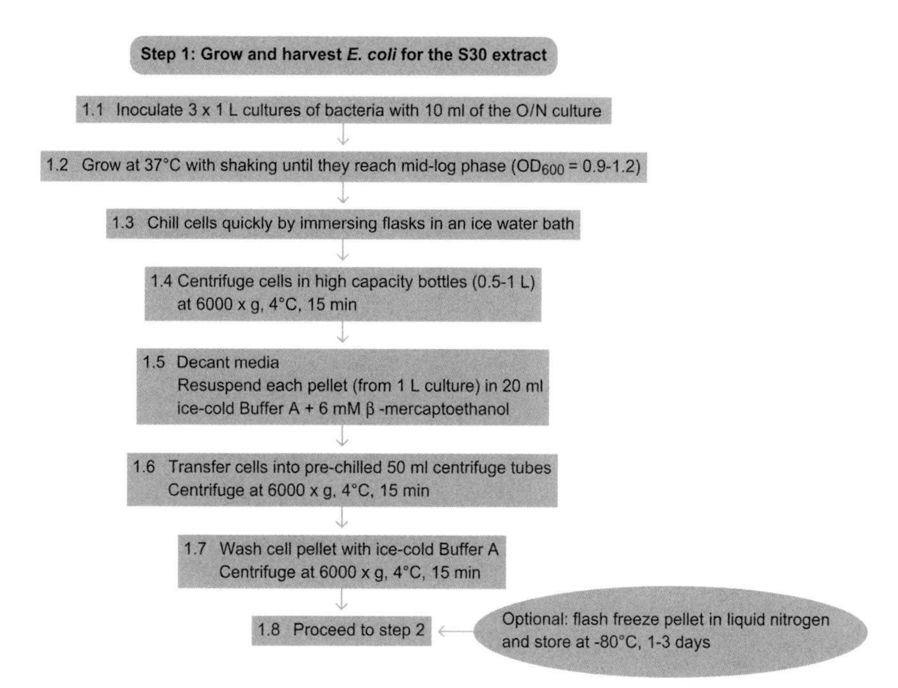

Figure 1.2 Flowchart of Step 1.

6. STEP 2 PREPARATION OF THE S30 EXTRACT

6.1. Overview

Crack the cells before removing cell debris by centrifugation. Activate the resulting extract to remove endogenous mRNA, and lastly, dialyze against a fresh buffer.

6.2. Duration

16 h

2.1 Carefully resuspend all of the cell pellets in 20 ml of ice cold Buffer A supplemented with 0.5 mM PMSF (add fresh), making sure to avoid frothing. Pass the cells through a syringe to ensure uniformity and to get rid of cell clumps that may clog the French press.

2.2 Crack the cells with a clean, prechilled French press using the manufacturer's instructions.

2.3 Centrifuge the lysate immediately at 30 000 × g, 4 °C for 30 min.

2.4 Decant the supernatant into a fresh centrifuge tube and repeat the centrifugation step. Discard the pellet.

2.5 Decant the clarified supernatant (amber in color) into a graduated cylinder to measure the final volume of the extract.

2.6 Add 0.3 times the volume of Activation Buffer. Transfer into a 50-ml polypropylene tube, cap, and incubate the extract at 37 °C for 80 min.

2.7 Dialyze the extract against at least 1 l of Buffer A at 4 °C overnight. Change the dialysis buffer and dialyze for another 2 h.

2.8 The extract is now ready. It should be immediately dispensed into aliquots (25–50 μl depending on the volume of the *in vitro* translation reactions used) and quickly frozen in liquid nitrogen.

2.9 The extract should be stored at −80 °C and should keep for several years.

6.3. Tip

E. coli *can be disrupted using other methods if a French press is not available (e.g., sonication or homogenization using glass beads).*

6.4. Caution

The French press exerts a high amount of pressure; therefore, make sure you receive proper training before using it.

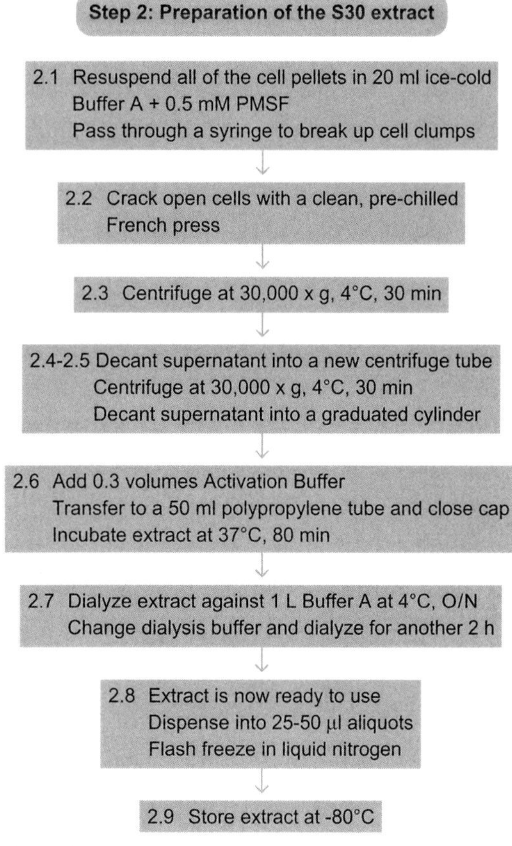

Figure 1.3 Flowchart of Step 2.

6.5. Caution

During the centrifugation step, make sure that the tubes are accurately balanced.

See Fig. 1.3 for the flowchart of Step 2.

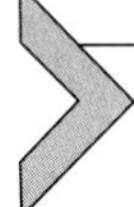

7. STEP 3 OPTIMIZATION OF THE COUPLED TRANSCRIPTION AND TRANSLATION REACTION

7.1. Overview

The optimal magnesium concentration for each S30 preparation must be empirically determined. Incubate the DNA of interest with the appropriate promoter in the presence of a radiolabeled amino acid under varied magnesium concentrations. Determine the total radioactivity incorporated into peptides using TCA precipitation (see TCA Precipitation) or SDS-PAGE analysis (see One-dimensional SDS-Polyacrylamide Gel Electrophoresis (1D SDS-PAGE)).

7.2. Duration

2–4 h

3.1 Prepare the DNA template with an *E. coli* promoter and ribosomal binding site upstream of the open reading frame sequence using standard molecular biology techniques (see Molecular Cloning, or Restrictionless cloning, or Rapid creation of stable mammalian cell lines for regulated expression of proteins using the Gateway® Recombination Cloning Technology and Flp-In T-REx® lines). A template with a T7 promoter can be used as long as T7 RNA polymerase is added to the incubation mixture to a concentration of 93 μg ml^{-1}.

3.2 Add to a 1.5-ml microcentrifuge tube (make up separate reactions with varying amounts of magnesium acetate):

DNA template	0.1–0.5 μg
S30 cocktail	15 μl
19 amino acids mixture	5 μl
Radiolabeled amino acid	up to 25 μCi
Magnesium acetate	5–15 mM
S30 extract	15 μl
T7 RNA polymerase	93 μg ml^{-1} (optional)
Sterile water	add to 50 μl final volume

3.3 Incubate at 37 °C for 30 min.

3.4 Hydrolyze peptidyl tRNA by adding KOH to a final concentration of 100 mM.

3.5 Stop the reaction by chilling on ice. For determination of the amount of the radioactive amino acid that was incorporated into peptides using TCA precipitation, follow Steps 3.6–3.10. If using SDS PAGE (e.g., if [^{35}S]-isotope is used), follow Steps 3.11–3.16.

3.6 Remove 10 μl of the translation reaction to a 1.5-ml microcentrifuge tube, add TCA to a final concentration of 5%, and incubate on ice for several minutes. BSA can be used as carrier by adding it to a final concentration of 0.5%.

3.7 Collect the acid precipitate by filtration on a glass fiber filter. Wash the filter 3 times with ice-cold 5% TCA, followed by one wash with ice-cold ethanol.

3.8 Dry the filter completely and count the amount of radioactivity using a scintillation counter.

3.9 To determine total input of radioactivity, pipette 10 μl of the translation reaction directly onto the glass fiber filter, dry it, and count.

3.10 Determine the fraction of the radiolabeled amino acid that was incorporated at each magnesium concentration. Once the optimal magnesium concentration is determined, this amount of magnesium acetate can be added to the S30 cocktail.

3.11 SDS-PAGE analysis of the radiolabeled product is routinely used to analyze *in vitro* translated proteins.

3.12 Add 5 volumes of cold acetone to precipitate the polypeptides. This step removes PEG that may cause streaking during SDS-PAGE analysis.

3.13 Centrifuge at 16 000 × g, 4 °C for 10 min. Resuspend the pellet in 100 μl SDS loading buffer and boil for 5 min.

3.14 Load the samples on the appropriate percentage SDS polyacrylamide gel.

3.15 Fix the gel by soaking in 10% TCA for 10 min twice, washing with MilliQ water in between.

3.16 Dry the gel and analyze using autoradiography. A typical gel is shown in Fig. 1.4.

7.3. Tip

The coupled transcription and translation reaction can be carried out on either circular or linear DNA.

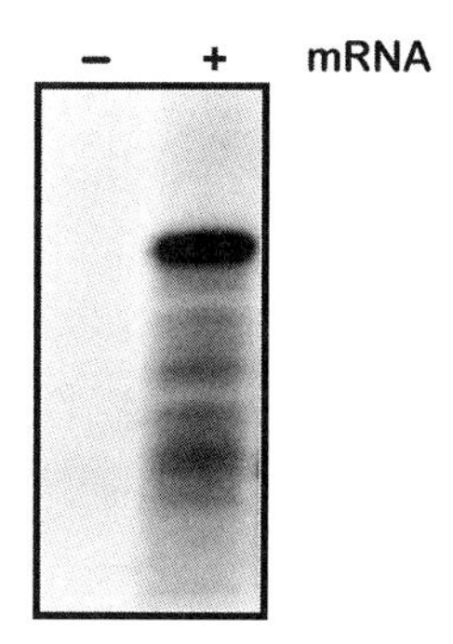

Figure 1.4 An SDS-PAGE gel of a typical *in vitro* translation assay using the procedure described here. In this case, mRNA rather than DNA was added to the reaction and ^{35}S-Methionine was added to the mixture to label the protein.

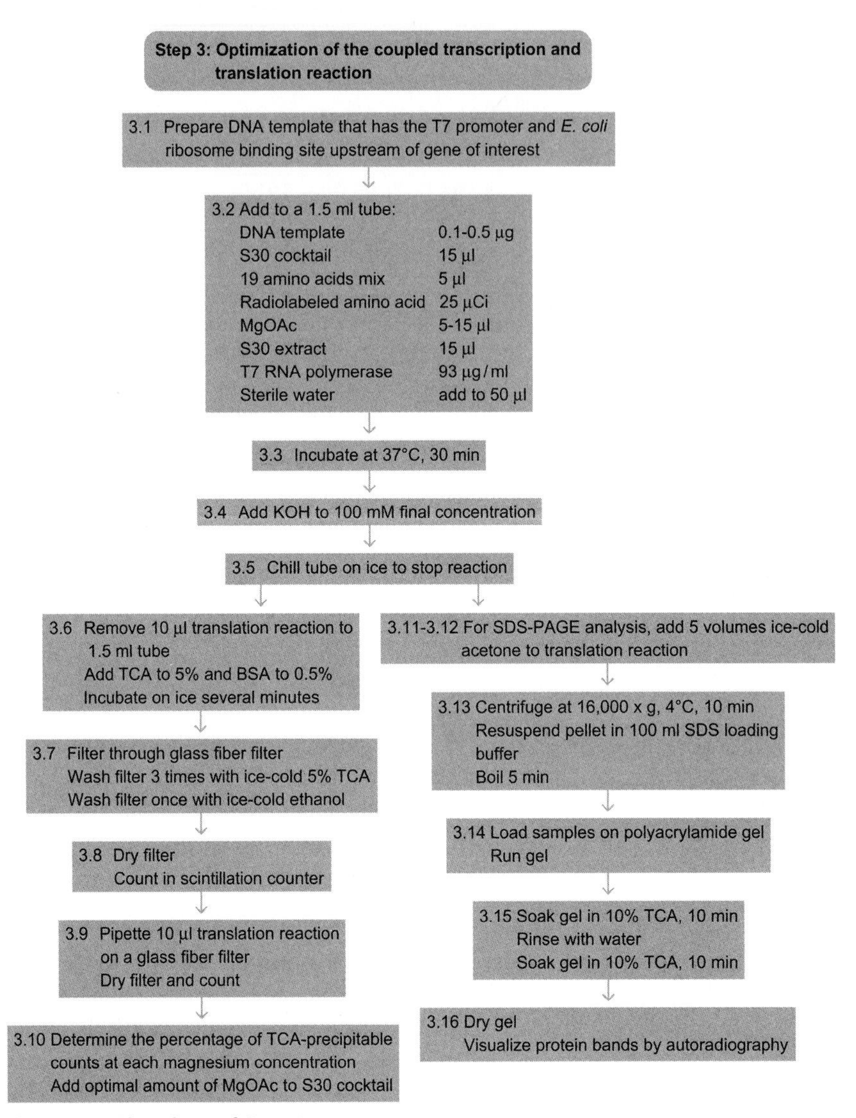

Figure 1.5 Flowchart of Step 3.

7.4. Tip

Although the coupled transcription translation reaction is recommended for increased protein yield, the procedure can be readily adapted for a translation reaction alone. In this case, in vitro transcribed RNA (5–20 μg) is added to the reaction instead of DNA.

See Fig. 1.5 for the flowchart of Step 3.

REFERENCES

Referenced Literature

Gong, F., & Yanofsky, C. (2001). Reproducing tna operon regulation in vitro in an S-30 system. Tryptophan induction inhibits cleavage of TnaC peptidyl-tRNA. *The Journal of Biological Chemistry*, *276*, 1974–1983.

Yokoyama, S., Hirota, H., Kigawa, T., et al. (2000). Structural genomics projects in Japan. *Nature Structural Biology*, *80* (supplement), 943–945.

Related Literature

Kigawa, T., Yabuki, T., Matsuda, N., et al. (2004). Preparation of *Escherichia coli* cell extract for highly productive cell-free protein expression. *Journal of Structural and Functional Genomics*, *5*, 63–68.

Mackie, G. A., Donly, B. C., & Wong, P. C. (1990). Coupled transcription-translation of ribosomal proteins. In G. Spedding (Ed.), *Ribosomes and Protein Synthesis a Practical Approach* (pp. 191–211). New York: Oxford University Press.

Referenced Protocols in Methods Navigator

One-dimensional SDS-Polyacrylamide Gel Electrophoresis (1D SDS-PAGE).

TCA Precipitation.

Molecular Cloning.

Restrictionless cloning.

Rapid creation of stable mammalian cell lines for regulated expression of proteins using the Gateway® Recombination Cloning Technology and Flp-In T-REx® lines.

CHAPTER TWO

Preparation of a *Saccharomyces cerevisiae* Cell-Free Extract for *In Vitro* Translation

Cheng Wu, Matthew S. Sachs[1]

Department of Biology, Texas A&M University, College Station, TX, USA

[1]Corresponding author: e-mail address: msachs@mail.bio.tamu.edu

Contents

Abstract

Eukaryotic cell-free *in vitro* translation systems have been in use since the 1970s. These systems can faithfully synthesize polypeptides when programmed with mRNA, enabling the production of polypeptides for analysis as well as permitting analyses of the *cis*- and *trans*-acting factors that regulate translation. Here we describe the preparation and use of cell-free translation systems from the yeast *Saccharomyces cerevisiae*.

Methods in Enzymology, Volume 539
ISSN 0076-6879
http://dx.doi.org/10.1016/B978-0-12-420120-0.00002-5

1. THEORY

Cell-free protein synthesizing systems from eukaryotes are in wide use in the identification of the components involved in translation. These systems are isolated from a variety of sources such as wheat germ, rabbit reticulocytes, ascites tumor cells, and fungi, including *S. cerevisiae*. Extracts from the first two sources are commercially available. However, there are several advantages of the yeast cell-free system. The preparation is simple, consisting of four major steps: cell breakage, collecting a clarified supernatant, chromatography to remove small endogenous molecules such as amino acids, and nuclease-treatment to eliminate endogenous RNAs. Yeast extracts can be prepared from wild-type or mutant strains, and from cells in a variety of different physiological states. The genetic and biochemical tractability of yeast makes it ideal for studies of the functions of components of the translational machinery and the functions of specific mRNA sequences in translational control.

2. EQUIPMENT

Heater incubator-shaker
Heater incubator
Refrigerated high-speed centrifuge
Sorvall GSA rotor (or equivalent)
Sorvall SS34 rotor (or equivalent)
−80 °C freezer
Drying oven
UV/vis spectrophotometer
Benchtop microcentrifuge
Polyacrylamide gel electrophoresis equipment
Micropipettors
500-ml Erlenmeyer flask
2-l Erlenmeyer flask
250-ml centrifuge bottles
50-ml polycarbonate centrifuge tubes
50-ml conical screw cap centrifuge tubes
2.5 × 10 cm flex columns
Micropipettor tips

Pasteur pipettes
Film developer
Autoradiography film
0.5-mm glass beads
1.5-ml microcentrifuge tubes
0.5-ml microcentrifuge tubes

3. MATERIALS

Yeast extract
Peptone
Glucose
HEPES
Potassium hydroxide (KOH)
Potassium acetate (KOAc)
Magnesium acetate (MgOAc)
Dithiothreitol (DTT)
Diethylpyrocarbonate (DEPC)
Mannitol
Phenylmethylsulfonyl fluoride (PMSF)
Sephadex G-25 superfine (Sigma G2550)
Micrococcal nuclease
Calcium chloride ($CaCl_2$)
EGTA
ATP
GTP
Phosphocreatine
Creatine phosphokinase
RNasin Plus RNase Inhibitor (Promega)
[^{35}S]-Methionine (>1000 Ci mmol^{-1}) (MP Biomedicals)
Dual-Luciferase Reporter Assay System (Promega)
Tris base
SDS
Glycerol
SDS polyacrylamide gel electrophoresis materials
Bromophenol blue
Liquid nitrogen

3.1. Solutions & buffers

Step 1 YPD growth medium

Component	Final concentration	Amount/l
Yeast extract	1%	10 g
Peptone	2%	20 g
Glucose	2%	20 g

Add water to 1 l, and divide into 100 ml (in 500-ml Erlenmeyer flask) and 800 ml (in 2-l Erlenmeyer flask) and autoclave

Buffer A

Component	Final concentration	Stock	Amount/l
HEPES–KOH, pH 7.6	30 mM	1 M	30 ml
KOAc, pH 7.0	100 mM	2 M	50 ml
MgOAc, pH 7.0	3 mM	0.3 M	10 ml
DTT	2 mM	2 M	1 ml

Add DEPC-treated water to 1 l, mix, and store at 4 °C

100 mM PMSF

Dissolve 1 g PMSF in 57.4 ml isopropanol, and store at room temperature

25 U μl^{-1} micrococcal nuclease

Dissolve 15000 U of micrococcal nuclease in 600 μl of DEPC-treated water, aliquot in 20 μl, freeze in liquid nitrogen, and store at −80 °C

0.1 M $CaCl_2$

Dilute from 1 M $CaCl_2$ stock with DEPC-treated water, store at 4 °C

0.17 M EGTA

Dilute from 0.5 M EGTA stock with DEPC-treated water and store at 4 °C

Step 2 10× Energy mix

Component	**Final concentration**	**Stock**	**Amount/ml**
HEPES–KOH, pH 7.6	200 mM	1 M	200 μl
ATP	10 mM	100 mM	100 μl
GTP	1 mM	100 mM	10 μl
Creatine phosphate	200 mM	500 mM	400 μl
DTT	20 mM	2 M	10 μl

Add DEPC-treated water to 1 ml, mix, aliquot in 200 μl, freeze in liquid nitrogen, and store at −80 °C

10 U μl^{-1} creatine phosphokinase

Component	**Final concentration**	**Stock**	**Amount/0.35 ml**
HEPES–KOH, pH 7.6	10 mM	1 M	3.5 μl
KOAc	50 mM	2 M	87.5 μl
Creatine phosphokinase	10 U μl^{-1}		3500 U
Glycerol	50%		175 μl

Add DEPC-treated water to 350 μl, mix, aliquot in 10 μl, freeze in liquid nitrogen, and store at −80 °C

0.1 M MgOAc, pH 7.0

Dilute from 0.3 M MgOAc stock with DEPC-treated water, and store at 4 °C

1 mM amino acid mix

Component	**Final concentration**	**Stock**	**Amount/ml**
20 individual amino acids	1 mM	100 mM	10 μl

Add DEPC-treated water to 1 ml, mix, freeze in liquid nitrogen, and store at −80 °C

1 mM amino acid mix (-Methionine)

Component	**Final concentration**	**Stock**	**Amount/ml**
19 individual amino acids	1 mM	100 mM	10 μl

Add DEPC-treated water to 1 ml, mix, freeze in liquid nitrogen, and store at −80 °C

4× SDS-PAGE loading buffer

Component	Final concentration	Stock	Amount/10 ml
Tris–HCl, pH 6.8	200 mM	1 M	2 ml
SDS	8%		0.8 g
Glycerol	40%		4 g
Bromophenol blue	0.5%		0.05 g
DTT	400 mM	2 M	2 ml

Add sterile deionized water to 10 ml

4. PROTOCOL

4.1. Duration

Preparation	About 1–2 days
Protocol	About 2 days

4.2. Preparation

Streak out *S. cerevisiae* cells from a frozen stock onto an YPD plate. Incubate at 30 °C for 1–2 days to obtain single colonies.

Sterilize the centrifuge bottles, tubes, and 1.5-ml microcentrifuge tubes by autoclaving. Sterilize the Pasteur pipettes and the glass beads by baking at 180 °C for 3 h.

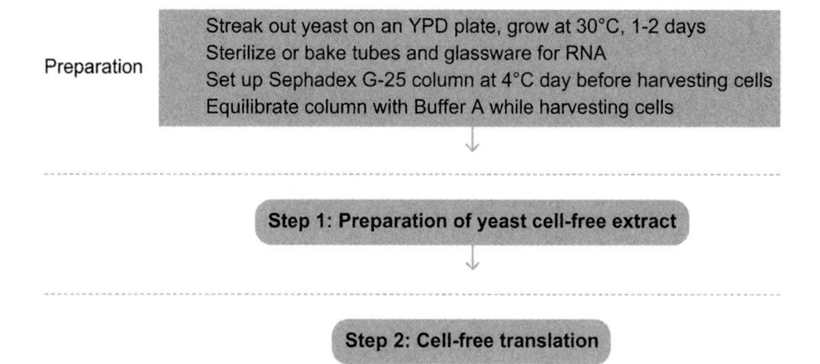

Figure 2.1 Flowchart for the complete protocol, including preparation.

In a 4 °C cold room, set up a 2.5 × 8 cm Sephadex G-25 column one day before breaking the cells. Preequilibrate the column with 250–300 ml of Buffer A plus 0.5 mM PMSF while harvesting the cells.

See Fig. 2.1 for the flowchart of the complete protocol.

5. STEP 1 PREPARATION OF YEAST CELL-FREE EXTRACT

5.1. Overview

Accomplish manual breakage of cells, centrifugation, and gravity chromatography of the extract to prepare clarified cell-free extract to be used in *in vitro* translation.

5.2. Duration

2 days

1.1 Inoculate a single colony into 100 ml of YPD growth medium.

1.2 Incubate at 30 °C with orbital shaking (200 rpm) for 24 h.

1.3 Measure OD_{600} of the starting culture.

1.4 Use starting culture to inoculate 800 ml of prewarmed YPD growth medium at an initial OD_{600}=0.03–0.06.

1.5 Incubate at 30 °C with orbital shaking (200 rpm) until $OD_{600} \approx 1.5$ (typical growth time is ~12–16 h).

1.6 Chill cells on ice. Carry out all the remaining steps for the preparation of the cell-free extract on ice or at 4 °C, except where noted.

1.7 Harvest cells by centrifugation at 3000 rpm in four centrifuge bottles in a Sorvall GSA rotor for 5 min at 4 °C. Decant and discard the supernatant.

1.8 Resuspend the cells by gentle mixing in a total volume of 30 ml of ice-cold Buffer A supplemented with 8.5% (w/v) mannitol. Transfer the resuspended cells to two preweighed 50-ml polycarbonate centrifuge tubes.

1.9 Centrifuge at 3000 rpm in a Sorvall SS34 rotor for 5 min at 4 °C. Decant and discard the supernatant.

1.10 Repeat Steps 1.8 and 1.9 four additional times with 10 ml of ice-cold Buffer A plus 8.5% mannitol for each tube for each wash. The final centrifugation is at 4000 rpm.

1.11 Determine the wet weight of the cells by weighing the tubes and subtracting the weight of the original empty tube.

1.12 Resuspend cells in 1.5 ml g^{-1} wet weight of Buffer A/8.5% mannitol/0.5 mM PMSF. Add six times cell weight of cold 0.5 mm glass beads.

1.13 In cold room, break cells by manual shaking for five 1 min periods with 1 min cooling on ice between shaking periods. Shaking is performed at a rate of 2 cycles s^{-1} over a 50 cm hand path. See Video 2.1, http://dx.doi.org/10.1016/B978-0-12-420120-0.00002-5.

1.14 Centrifuge at 1000 rpm in the Sorvall SS34 rotor for 2 min at 4 °C to collect the glass beads. Transfer the supernatant with a sterile Pasteur pipette to the fresh polycarbonate centrifuge tubes.

1.15 Centrifuge the broken cells at 16000 rpm in the Sorvall SS34 rotor for 6 min at 4 °C.

1.16 Collect the supernatant with a sterile Pasteur pipette, avoiding both the pellet and the fatty upper layer, and transfer the supernatant to a fresh 50-ml conical tube that is maintained on ice. Repeat Step 1.15 once more if the supernatant is not clear enough or some loose pellet is accidentally picked up.

1.17 Chromatograph the clarified supernatant by gravity flow on the G-25 column. See Video 2.2, http://dx.doi.org/10.1016/B978-0-12-420120-0.00002-5.

1.18 Collect fractions (~0.5 ml) and measure the absorbance at 260 nm, using a spectrophotometer. Pool the void-volume fractions containing the A_{260} peak and neighboring fractions whose A_{260} are 75% of the peak value in a 50-ml conical tube maintained on ice.

1.19 Add $CaCl_2$ to a final concentration of 1 mM, micrococcal nuclease to 50 U ml^{-1}, incubate at 26 °C for 10 min, stop the nuclease treatment by adding EGTA to 2.5 mM final concentration.

1.20 Freeze extract directly in liquid nitrogen after preparing 100 μl aliquots. Store at −80 °C.

5.3. Tip

Add DTT and PMSF to Buffer A immediately prior to use.

5.4. Tip

For larger scale preparation, a bigger column is needed, for example, 2.5 × 20 cm.

See Fig. 2.2 for the flowchart of Step 1.

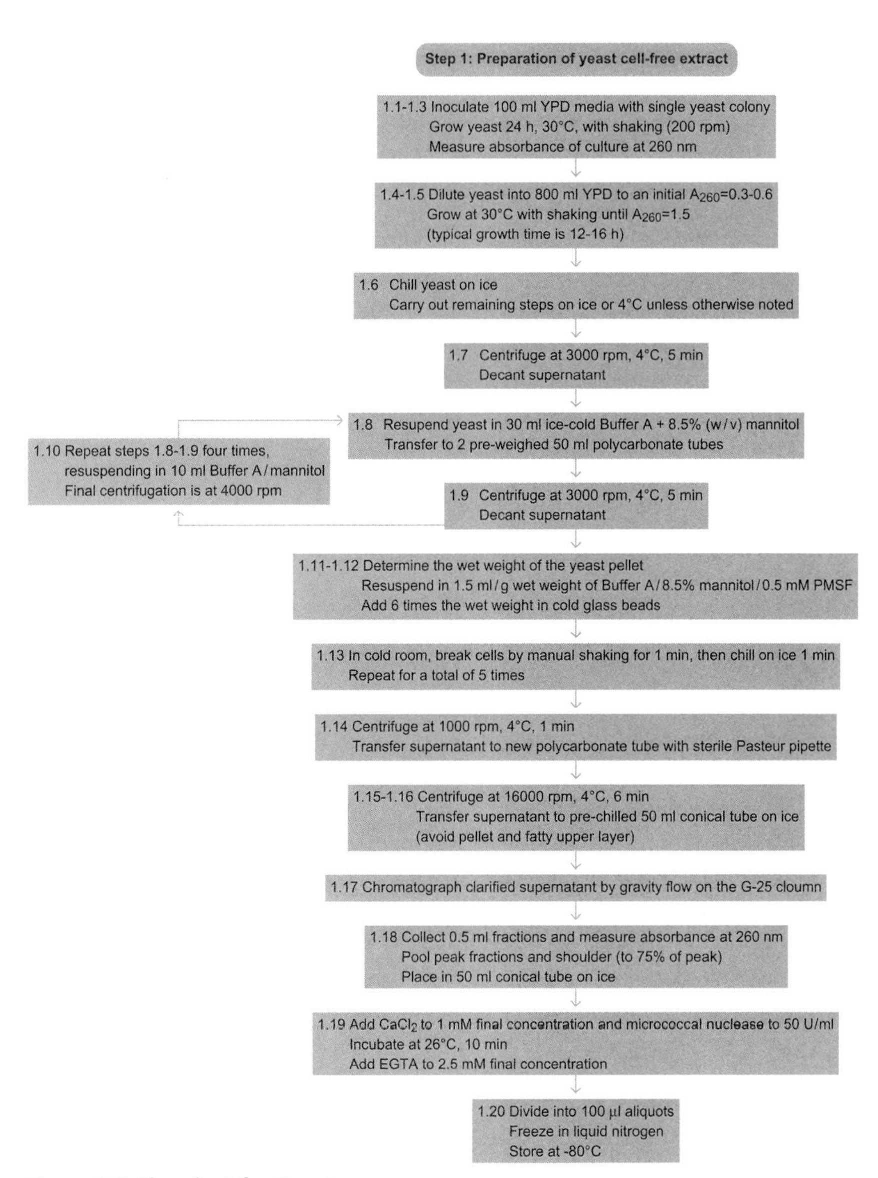

Figure 2.2 Flowchart for Step 1.

6. STEP 2 CELL-FREE TRANSLATION

6.1. Overview

Perform an *in vitro* translation using the cell-free extract.

6.2. Caution

Consult your institute Radiation Safety Officer for proper ordering, handling, and disposal of radioactive materials.

6.3. Duration

1 h

2.1 Thaw reagents, buffers, the cell-free extract and mRNA samples on ice. The following procedure is to accomplish 20 10-μl reactions.

2.2a To perform a dual luciferase assay, in a 0.5-ml microcentrifuge tube, assemble a master solution by adding 10 μl of 10× energy mix, 0.6 μl of creatine phosphokinase, 2.5 μl of 2 M KOAc, 2 μl of 0.1 M MgOAc, 1 μl of 1 mM amino acids mix, 1 μl of RNasin, mRNA specifying *Renilla* luciferase, and DEPC-treated water to a final volume of 80 μl. Gently mix, and keep on ice.

2.2b To radiolabel the translation products with [^{35} S]-Methionine, in a 0.5-ml microcentrifuge tube assemble a master solution by adding 10 μl of 10× energy mix, 0.6 μl of creatine phosphokinase, 2.5 μl of 2 M KOAc, 2 μl of 0.1 M MgOAc, 1 μl of 1 mM amino acids mix (-Met), 1 μl of RNasin, 500 μCi of [^{35} S]-Methionine, and DEPC-treated water to a final volume of 80 μl. Mix gently, keep on ice.

2.3 Aliquot 4 μl of the master solution to each individual reaction tube.

2.4 Add 1 μl of appropriately diluted mRNA (e.g., synthetic mRNA specifying firefly luciferase).

2.5 Add 5 μl of the cell-free extract and mix gently.

2.6 Incubate reactions at 26 °C (typically for 15–30 min).

2.7a A dual luciferase assay can be performed here. If the Dual-Luciferase Reporter Assay System from Promega is used, follow the instructions on the technical manual (http://www.promega.com/tbs/tm040/tm040.pdf), first stopping reactions by adding 5× passive lysis buffer to a final concentration of 1×.

2.7b To examine radiolabeled translation products on a gel, stop reactions by adding 4× SDS-PAGE loading buffer to a final concentration of 1×.

Load the samples on a polyacrylamide gel, separate the proteins by SDS-PAGE (see One-dimensional SDS-Polyacrylamide Gel Electrophoresis (1D SDS-PAGE)), fix and dry gel, and visualize by autoradiography.

6.4. Tip

The Renilla luciferase mRNA added in Step 2.2a is used as an internal control in the dual luciferase assay.

6.5. Tip

The amount of mRNA used to radiolabel the translation products is usually tenfold higher than is used to perform the dual luciferase assay (e.g., 60 ng vs. 6 ng of synthetic mRNA).

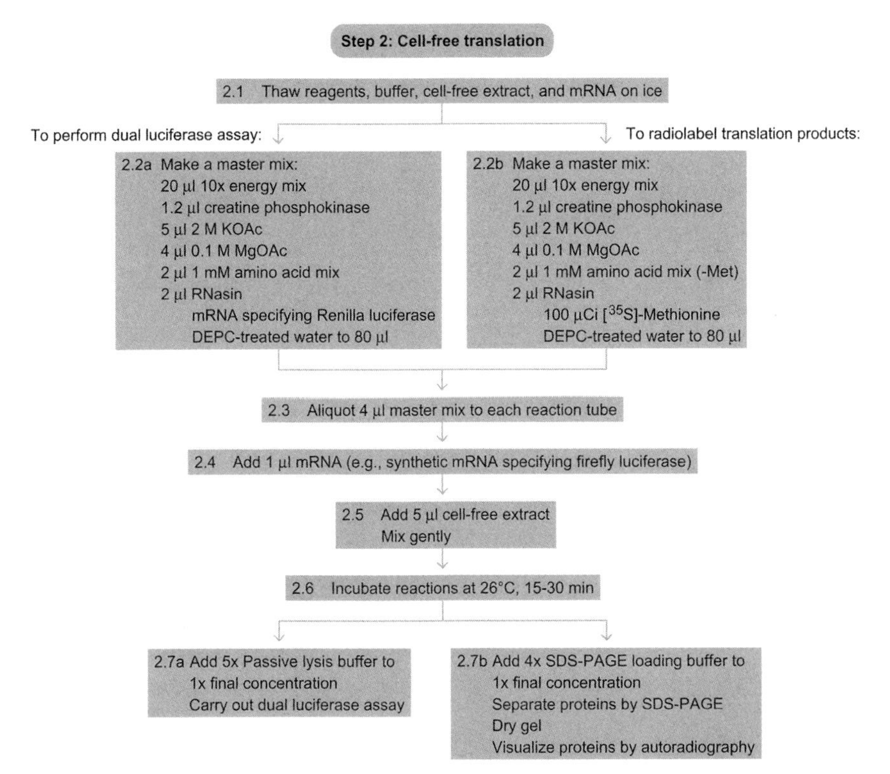

Figure 2.3 Flowchart for Step 2.

6.6. Tip

The volume of an individual translation reaction can be increased as needed.

6.7. Tip

For each individual batch of cell-free extract, titration of K^+ and Mg^{2+} may be needed to optimize translation. The amounts given in this protocol can be viewed as a starting point. Different mRNAs may also have different ionic optima for translation.

6.8. Tip

After adding passive lysis buffer, translation products can be kept at room temperature for at least 1 h.

See Fig. 2.3 for the flowchart of Step 2.

SOURCE REFERENCES

Wu, C., Amrani, N., Jacobson, A, & Sachs, M. S. (2007). The use of fungal in vitro systems for studying translational regulation. *Methods in Enzymology*, *429*, 203–225.

Related Literature

Tarun, S. Z., Jr., & Sachs, A. B. (1995). A common function for mRNA 5′ and 3′ ends in translation initiation in yeast. *Genes & Development*, *9*(23), 2997–3007.

Wang, Z., & Sachs, M. S. (1997). Arginine-specific regulation mediated by the *Neurospora crassa arg-2* upstream open reading frame in a homologous, cell-free in vitro translation system. *The Journal of Biological Chemistry*, *272*(1), 255–261.

Referenced Protocols in Methods Navigator

One-dimensional SDS-Polyacrylamide Gel Electrophoresis (1D SDS-PAGE).

SECTION II

Protein Protocols/ Protein *In Vivo* Binding Assays

CHAPTER THREE

Yeast Two-Hybrid Screen

Lauren Makuch[1]
Department of Neuroscience, Johns Hopkins University School of Medicine, Baltimore, MD, USA
[1]Corresponding author: e-mail address: lmakuch1@jhmi.edu

Contents

Methods in Enzymology, Volume 539
ISSN 0076-6879
http://dx.doi.org/10.1016/B978-0-12-420120-0.00003-7

Abstract

Yeast two-hybrid is a method for screening large numbers of gene products (encoded by cDNA libraries) for their ability to interact with a protein of interest. This system can also be used for characterizing and manipulating candidate protein: protein interactions. Interactions between proteins are monitored by the growth of yeast plated on selective media.

1. THEORY

The two-hybrid system is an *in vivo* yeast-based system that takes advantage of the modular nature of the yeast GAL4 transcription factor. GAL4 has two domains whose activities can be separated: the **DNA**-**B**inding domain and the transcriptional **A**ctivation **D**omain. The two-hybrid system identifies the interaction between two proteins (X and Y) by reconstituting these two GAL4 domains and thus allowing transcriptional activation of a reporter gene, which has been designed to be a selectable marker. Reconstitution of the DNA-binding domain and the activation domain will only occur if X and Y interact. To achieve this, two fusion proteins are constructed, one of which contains the DNA-binding domain fused to the first protein of interest (DB-X, also called the 'bait') and the other of which contains the activation domain fused to the second protein of interest (AD-Y, also called the 'prey'). DB-X–AD-Y interaction reconstitutes a functional transcription factor that activates reporter genes driven by promoters containing the relevant DB sites. A selectable marker such as *HIS3* is used as a reporter gene, and transcription activation resulting from the interaction of DB-X and AD-Y can therefore be assessed by a growth of cells on agar plates lacking histidine. Thus, the yeast two-hybrid system allows for the detection of protein–protein interactions genetically.

2. EQUIPMENT

Water baths (30 and 42 °C)
Incubator (30 °C for yeast, 37 °C for *E. coli*)
Autoclave
UV/Vis spectrophotometer
Centrifuge (capable of spinning 250-ml bottles at 3000 g)
Shaking incubator (30 °C for yeast, 37 °C for *E. coli*)

Electroporator
Heat block (95–100 °C)
Vacuum aspirator
Microcentrifuge
Micropipettors
Micropipettor tips
250-ml centrifuge bottles
Magnetic stir bars
10-cm Petri plates
15-cm Petri plates
Sterile loop (autoclaved)
1.5-ml microcentrifuge tubes
1.5-ml microcentrifuge tubes (with locking cap)
0.5-mm glass beads, soaked in ethanol and then autoclaved
425–600-μm glass beads (separate use)
Disposable cuvettes (1 ml)
0.2-μm filter units
Electroporation cuvettes
Sterile 14-ml snap-cap tubes

3. MATERIALS

cDNA library of choice, cloned into pPC86 (commercially available)
Appropriate yeast strain (PJ69, CG1945, or Y190)
pPC86 DNA (Invitrogen)
pDBLeu DNA (Invitrogen)
YPAD medium (rich medium for routine growth of yeast)
Synthetic complete medium (SC)
3-Amino-1,2,4-triazole (3AT)
Luria broth (LB)
Autoclaved, distilled water
Tris base
Hydrochloric acid (HCl)
EDTA
Sonicated herring or salmon sperm DNA (10 mg ml^{-1})
Lithium acetate (LiOAc)
Polyethylene glycol-3350 (PEG-3350)
Chemically competent Top10 or DH5α *E. coli*

Electrocompetent *E. coli* cells
Kanamycin
Ampicillin
Phenol
Chloroform
Isoamyl alcohol
Ethanol
Triton X-100
Sodium dodecyl sulfate (SDS)
Sodium chloride (NaCl)

3.1. Solutions & Buffers

For all yeast and bacterial media recipes, please refer to *Saccharomyces cerevisiae* Growth Media, Growth Media for *E. coli*, and Pouring Agar Plates and Streaking or Spreading to Isolate Individual Colonies.

Step 1 2 M 3-Amino-1,2,4-triazole (3AT)

Dissolve 42.04-g 3AT in 250-ml purified water. Filter through a 0.2-μm filter unit.

50% (w/v) PEG-3350

Add 35-ml ddH_2O to 50-g PEG 3350 in a 150-ml glass beaker. Stir with a magnetic stir bar until dissolved. Transfer all of this mixture to a 100-ml graduated cylinder. Rinse the beaker with a small amount of distilled water and add this to the graduated cylinder holding the PEG solution. Add ddH_2O until the volume is exactly 100 ml. Mix the solution thoroughly by inversion and transfer all of it to a tightly capped container that can be autoclaved. Sterilize the PEG solution by autoclaving it.

10× LiAc

Dissolve 10.2-g LiAc in purified water in a final volume of 100 ml to make a 1-M stock solution.

10× TE

Component	Final concentration	Stock	Amount
Tris–HCl, pH 7.5	100 mM	500 mM	20 ml
EDTA	100 mM	500 mM	2 ml

Add purified water to 100 ml. Sterilize by autoclaving.

Step 4 Buffer A

Component	Final concentration	Stock	Amount
Tris–HCl, pH 8.0	10 mM	1 M	0.4 ml
NaCl	100 mM	5 M	0.8 ml
EDTA	1 mM	500 mM	80 μl
Triton X-100	2% (w/v)	20% (w/v)	4 ml
SDS	1% (w/v)	20% (w/v)	2 ml

Add purified water to 40 ml and stir to completely dissolve.

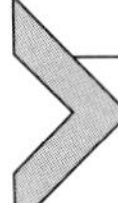

4. PROTOCOL

4.1. Preparation

Due to the complexity of this procedure, it is strongly advised that you read the protocol in its entirety before beginning a two-hybrid screen. Prior to starting a two-hybrid screen, determine as much information as possible regarding the 'bait' protein of interest (X) and those that might be identified as interactors (Y). From the ProQuest Two Hybrid Instruction Manual, several questions that are of particular concern are

- Does the bait protein function as a transcription activator or contain other domains of known function (repressor, etc.)?
- Do the bait or predicted prey proteins belong to a protein family?
- Where and when is the test protein expressed? It may be desirable to construct the cDNA library from this tissue during this developmental time period.
- How will the interactions between proteins identified in the two-hybrid screen be confirmed?

Cloning and transformation methods are used on the sections below. For more details on these procedures, please see Molecular Cloning, Explanatory chapter: PCR -Primer design and Transformation of Chemically Competent *E. coli*.

4.2. Construction of the DB fusion protein

Clone the bait DNA fragment into the GAL4 DNA-binding domain vector, pDBLeu. Make sure your protein will be translated in the same reading frame as the GAL4 DNA-binding domain. The cloning can either be done by digesting the bait DNA fragment with the same enzymes as the pDBLeu vector and then ligating them, or by PCR amplification of the bait DNA fragment, using primers designed to contain restriction sites in the appropriate positions to allow in-frame fusion.

Ligations should be transformed into a chemically competent *E. coli* strain and plated onto LB agar plates containing 25 $\mu g\ ml^{-1}$ kanamycin. Sequencing the vector/insert DNA junctions is highly recommended to confirm the reading frame.

4.3. Construction of the AD fusion protein

If testing the interaction between two known proteins, clone the prey DNA fragment (Y) into the GAL4 activation domain vector, pPC86. Make sure your protein will be translated in the same reading frame as the GAL4 activation domain. The cloning can either be done by digesting the prey DNA fragment with the same enzymes as the pPC86 vector and then ligating them, or by PCR amplification of the prey DNA fragment, using primers designed to contain restriction sites in the appropriate positions to allow in-frame fusion. Ligations should be transformed into a chemically competent *E. coli* strain and plated onto LB agar plates containing 100 $\mu g\ ml^{-1}$ ampicillin. Sequencing the vector/insert DNA junctions is highly recommended to confirm the reading frame.

4.4. Testing for Self-activation of pDBLeu-X (and pPC86-Y)

Prior to starting a two-hybrid screen, it is necessary to test pDBLeu-X (and pPC86-Y, if studying two known proteins) for self-activation and determine the basal expression level of the *HIS3* reporter gene. By increasing the amount of 3AT (from 0 to 100 mM 3AT) in yeast media plates lacking histidine, HIS3 activity can be titrated down to a point at which growth in the absence of histidine is inhibited. This point will vary depending on the individual bait as well as the yeast strain you are using.

HIS3 encodes an enzyme involved in histidine biosynthesis, called imidazole glycerol phosphate dehydratase. It can be inhibited in a dose-dependent manner by 3AT. Different yeast strains express variant levels of HIS3 basally (this is termed 'leakiness'). By including the minimum effective dose of 3AT (which will be determined in the following steps) in the yeast media plates used for the screen, you will increase the likelihood of identifying weak protein–protein interactions that would otherwise be masked by basal HIS3 expression. For best results, test pDBLeu-X in the presence of the AD vector because a DB-X–AD complex could form and activate transcription.

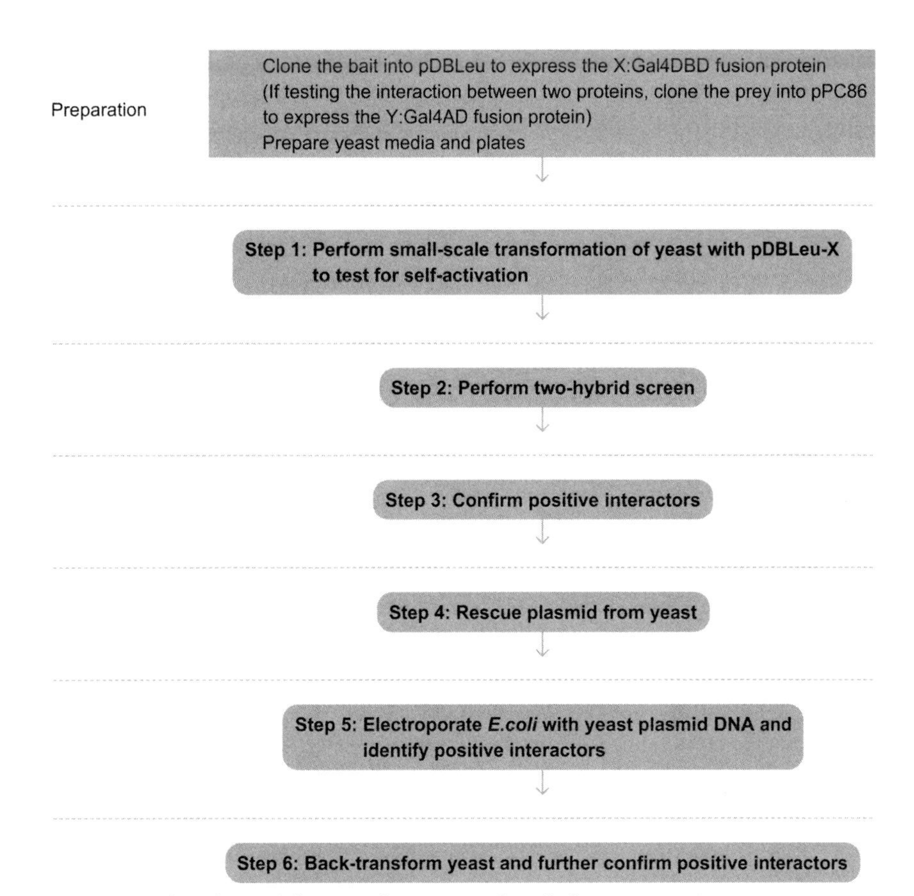

Figure 3.1 Flowchart of the complete protocol, including preparation.

4.5. Duration

Preparation	Variable
Protocol	2–3 weeks

See Fig. 3.1 for the flowchart of the complete protocol.

5. STEP 1 SMALL-SCALE TRANSFORMATION OF YEAST WITH pDBLeu-X

5.1. Overview

Before beginning the two-hybrid screen, it is necessary to transform yeast (for more information, see Chemical Transformation of Yeast) with the bait vector alone. When testing for self-activation, it is best to follow this protocol and to transform both the bait vector and the empty pPC86 vector. In this case, the total amount of DNA transformed should not exceed 3 μg.

5.2. Duration

~5 days

1.1 Prepare the stock solution of 3AT and of all amino acids. Refer to *Saccharomyces cerevisiae* Growth Media for the preparation of yeast media and related solutions.

1.2 Prepare agar plates. The following list indicates the minimum number of plates of each type that will be needed. The notation 'SC-Leu' or 'SC-Leu-Trp' indicates media lacking leucine or leucine and tryptophan, respectively.

2 YPAD
6 SC-Leu
18 SC-Leu-Trp
2 SC-Leu + 0 mM 3AT
2 SC-Leu + 10 mM 3AT
2 SC-Leu + 25 mM 3AT
2 SC-Leu + 50 mM 3AT
2 SC-Leu + 75 mM 3AT
2 SC-Leu + 100 mM 3AT
2 SC-Leu-His + 0 mM 3AT
2 SC-Leu-His + 10 mM 3AT
2 SC-Leu-His + 25 mM 3AT
2 SC-Leu-His + 50 mM 3AT
2 SC-Leu-His + 75 mM 3AT
2 SC-Leu-His + 100 mM 3AT
2 SC-Leu-Trp-His + 0 mM 3AT
2 SC-Leu-Trp-His + 10 mM 3AT
2 SC-Leu-Trp-His + 25 mM 3AT
2 SC-Leu-Trp-His + 50 mM 3AT
2 SC-Leu-Trp-His + 75 mM 3AT
2 SC-Leu-Trp-His + 100 mM 3AT

1.3 Prepare 100-ml YPAD broth in a 500-ml flask for the transformation.

1.4 Inoculate 5-ml YPAD media with a colony of yeast using a sterile inoculation loop and sterile technique. Briefly vortex to disperse the cells. Grow the culture overnight at 30 °C in a shaking incubator (shaking at 250 rpm).

1.5 The next morning, make a 1:5 or 1:10 dilution of your overnight culture into fresh prewarmed YPAD media. After dilution, there should be enough transferred cells to make the media slightly cloudy. Make sure you dilute into enough volume to have 2.5 ml of fresh culture per transformation (e.g., for five transformations, you will need about 12.5 ml of culture).

1.6 Remove the 50% PEG 3350 from the refrigerator and allow it to warm to room temperature on the bench top.

1.7 Grow the diluted culture another 4 h at 30 °C in a shaking incubator.

1.8 Spin down the cells at 3000 rpm for 10 min.

1.9 Denature 10 mg ml^{-1} sonicated salmon sperm DNA by incubating it at 95–100 °C for 10–15 min. Immediately put the tube on ice.

1.10 Using a vacuum aspirator, remove the supernatant from the yeast cultures and resuspend the cells in 1 ml of sterile, autoclaved water per transformation.

1.11 Aliquot 1 ml of cells into individual sterile microcentrifuge tubes.

1.12 Spin cells at 8000 rpm for 30 s.

1.13 Remove the supernatants. Then, **layer** onto the cell pellets the following reagents in this order:

120-μl 50% PEG 3350
18-μl 1 M LiAc
2.5-μl 10 mg ml^{-1} salmon sperm DNA
35-μl water plus 1–2 μg DNA (the volume of the DNA depends on the concentration of the pDBLeu-X. The remainder of the 35-μl volume should be made up with sterile water).

1.14 Vortex the tubes for at least 1 min to resuspend the cells. Sometimes, the cells will never be completely dispersed, but the procedure will still work.

1.15 Incubate the cells in a 30 °C water bath for 30 min.

1.16 Heat-shock the cells at 42 °C for 20 min and then let them sit on ice for 1 min.

1.17 Spin-down the tubes in a microcentrifuge at 8000 rpm for 30 s.

1.18 Remove the supernatants with a micropipettor. Resuspend the cells in 500 μl of sterile water by pipetting up and down. Plate 50 μl of cells onto selection media. In this case, plate the transformed yeast onto SC-Leu, SC-Leu-His, SC-Leu + 3AT, and SC-Leu-His + 3AT plates. Save the rest of the cells at 4 °C and incubate the plates inverted at 30 °C for 2–3 days.

1.19 After growing for 2–3 days, determine the concentration of 3AT needed to eliminate self-activation of your bait vector. Look for the concentration of 3AT at which growth occurs on SC-Leu + 3AT but is eliminated on SC-Leu-His + 3AT plates. Growth of yeast on SC-Leu containing this concentration of 3AT indicates that the 3AT is not eliminating growth due to toxicity.

1.20 IMPORTANT: Make 1 l of SC-Leu-Trp-His agar containing the appropriate concentration of 3AT and pour into 25 15-cm plates on which to conduct the two-hybrid screen. Also prepare 1 L of liquid YPAD media and 100 ml of SC-Leu liquid media.

See Fig. 3.2 for the flowchart of Step 1.

6. STEP 2 TWO-HYBRID SCREEN

6.1. Overview

After transforming yeast with pDBLeu-X and determining the concentration of 3AT needed for conducting the two-hybrid screen, it is necessary to grow up enough pDBLeu-X yeast for the large-scale screen. Subsequently, the pDBLeu-X yeast will be transformed with the pPC86-cDNA library to identify protein interactors.

6.2. Duration

1 week

2.1 Two days before starting the two-hybrid screen, make a 2–3-ml overnight culture of pDBLeu-X-transformed yeast in SC-Leu liquid media and grow it, shaking (250 rpm), at 30 °C.

2.2 The next day, warm 100 ml of SC-Leu liquid media to 30 °C.

2.3 Place 0.5–1 ml of the previous day's culture into the 100 ml of SC-Leu liquid media and grow, shaking, at 30 °C overnight. *Reading the remainder of this protocol prior to starting the two-hybrid screen the next day is highly recommended.*

2.4 The next morning, you will transform yeast already containing the bait vector with a cDNA library of your choice in the pPC86 vector. Before beginning, place 500 ml of YPAD liquid media in a 2-l flask in a 30 °C shaking incubator. Fill a cuvette with 750-μl dH_2O and 250 μl of the 100-ml overnight culture of transformed yeast.

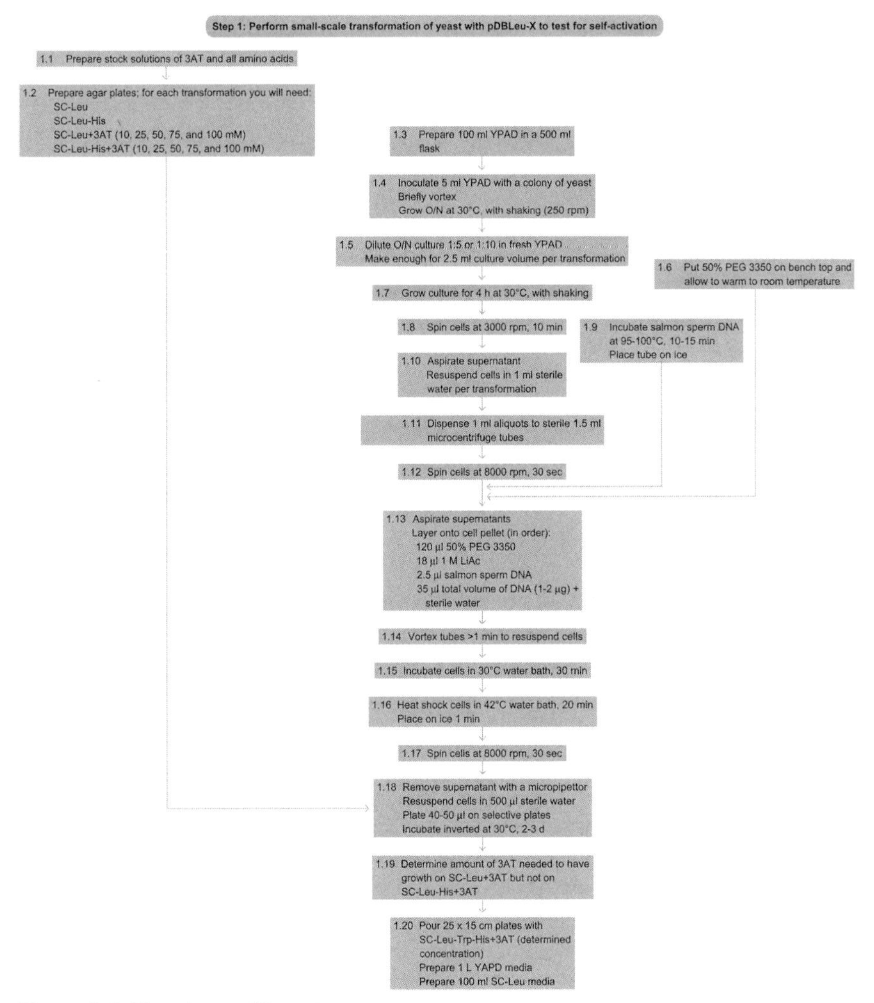

Figure 3.2 Flowchart of Step 1.

2.5 Using the spectrophotometer, measure the absorbance at 600 nm. Multiply this absorbance value by 4 since a 1:4 dilution was measured. The product of this multiplication has units of OD/ml.

2.6 Determine the volume of the 100-ml overnight culture required to give 50-OD units. For example, if you have 1 OD ml^{-1}, then you need 50 ml of your overnight culture to obtain 50 OD of cells.

2.7 Add 50 OD of your overnight culture to the 500-ml YPAD prewarmed to 30 °C. Continue growing the culture for 5 h in a 30 °C shaking incubator.

2.8 Remove 50% PEG-3350 from the refrigerator and allow it to reach room temperature on the bench top.

2.9 After the pDBLeu-X-transformed yeast has grown for 4.5 h, prepare the following reagents fresh:

(a) 1× TE/LiAc by combining:
11-ml 10× TE
11-ml 10× LiAc
88-ml autoclaved water

(b) PEG/LiAc by combining:
2-ml 10× TE
2-ml 10× LiAc
16-ml 50% PEG-3350

(c) 250-μl carrier DNA by boiling sonicated herring or salmon sperm DNA (10 mg ml^{-1}) for 5 min, and placing on ice until use

2.10 Split the yeast cells evenly into three 250-ml centrifuge tubes and centrifuge at 3000 × *g* for 5 min at room temperature.

2.11 Pour off the supernatants and resuspend each pellet by pipetting up and down in 50-ml autoclaved, distilled water. Combine one centrifuge bottle with the other two so you now have only two centrifuge bottles containing all the cells.

Spin at 3000 × *g* for 5 min at room temperature.

2.12 Decant the supernatants and suspend each cell pellet in 50-ml 1× TE/LiAc.

2.13 Spin at 3000 g for 5 min at room temperature.

2.14 Very gently pour off the supernatants and suspend each pellet in a final volume of 1.25-ml 1× TE/LiAc solution and pool both suspensions for a total volume of 2.5–3 ml. Transfer all the cells to a 50-ml conical tube.

2.15 Perform 25 transformations. Combine all the cells (2.5–3 ml) with 250-μl freshly boiled carrier DNA and 50 μg of pPC86-cDNA library. Mix gently, but completely, by pipetting up and down. Add 18-ml PEG/ LiAc solution and mix gently by pipetting up and down. Aliquot into 25 autoclaved microcentrifuge tubes of 700–800 μl each.

2.16 Incubate for 30 min in a 30 °C water bath.

2.17 Heat-shock for 15 min in a 42 °C water bath.

2.18 Spin the tubes in a microcentrifuge at 6000–8000 × *g* for 20–30 s at room temperature. Remove the supernatants taking care not to disrupt the cell pellets. Gently suspend each pellet in 400-μl autoclaved distilled water by pipetting up and down.

2.19 As time allows, it is necessary to titrate the transformation of the cDNA library to determine how well the full library was represented (to measure approximately how many *different* gene products were sampled).

(a) From one tube of the 25 tubes, plate three dilutions of the transformation. Mix 10-μl transformation with 90-μl autoclaved distilled water. Plate the entire 100 μl on a 10-cm SC-Leu-Trp plate (1:40 final dilution factor).

(b) From the same transformation tube, mix 10-μl transformation and 990-μl autoclaved water. Plate 100 μl on a 10-cm SC-Leu-Trp plate (1:400 final dilution factor).

(c) From the dilution in Step (b) above, add 100-μl–900-μl autoclaved water. Plate 100 μl of this mixture on a 10-cm SC-Leu-Trp plate (1:4000 final dilution factor).

2.20 Plate the entire transformation mixture from one of the 25 transformations onto a single 15-cm SC-Leu-Trp-His + 3AT plate. Repeat for the remaining 24 transformations. Spread the transformation mixtures onto the plates by pouring a few (~10) sterile, dry glass beads onto each plate and then shaking the plates vigorously. When the mixtures have been spread evenly over all the plates, pour the dirty glass beads into a beaker of ethanol where they can be cleaned.

2.21 Incubate all plates (including titration plates) at 30 °C for 3 days.

2.22 Calculate the number of transformants on the 10-cm titration plates as follows:

(a) Count the number of colonies on the plates having 20–300 colonies.

(b) Multiply the number of colonies counted by the dilution factor. For example, if you count 25 colonies on the 1:4000 plate, $25 \times 4000 = 1 \times 10^5$ colonies per tube.

See Fig. 3.3 for the flowchart of Step 2.

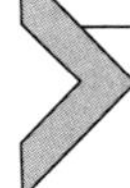

7. STEP 3 CONFIRMATION OF POSITIVE INTERACTORS

7.1. Overview

The appropriate method for confirming positive interactors depends on the strain of yeast in which the screen has been conducted. The following procedure describes the confirmation of interactors in PJ69, where the second

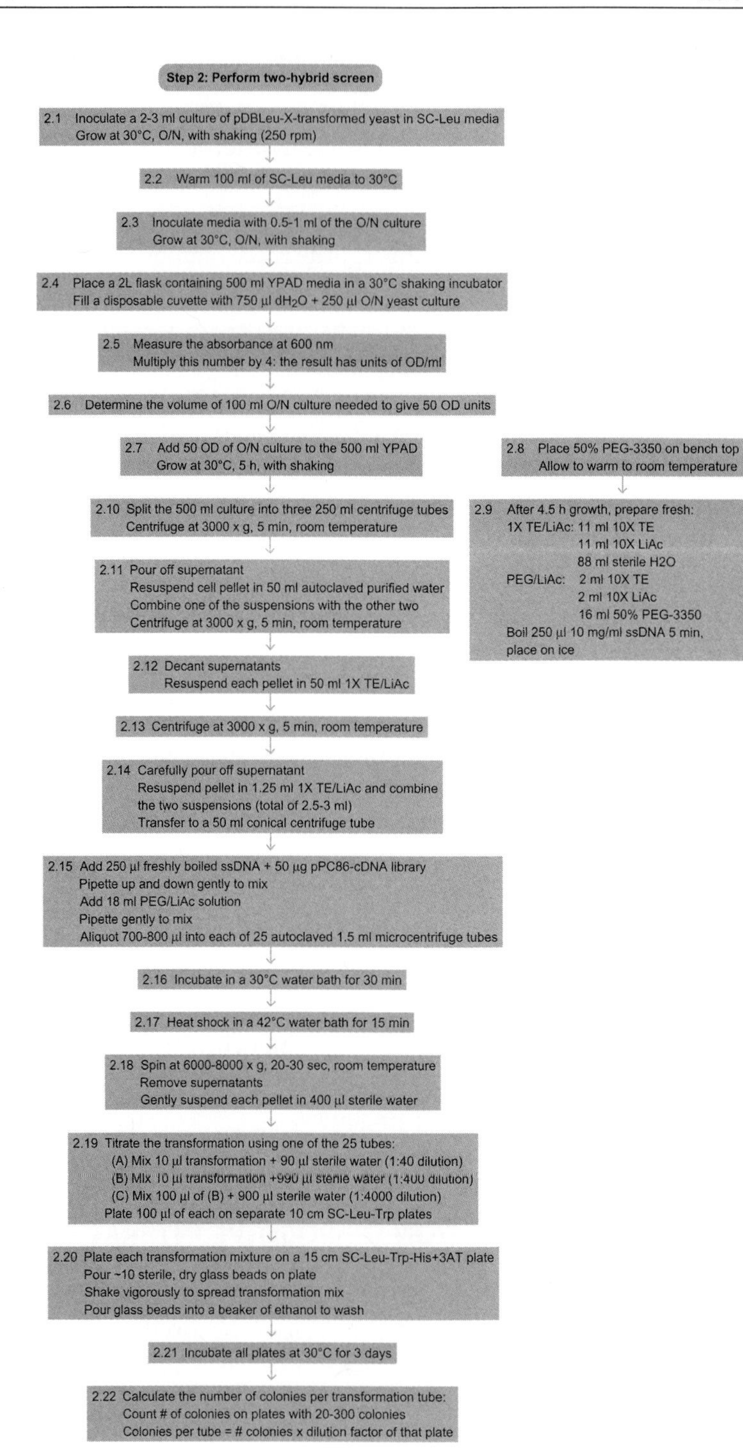

Figure 3.3 Flowchart of Step 2.

selection is adenine. If the screen was conducted in CG1945 or Y190, the second selection will be expression of β-galactosidase. In this case, replica plating of positive yeast colonies onto a nitrocellulose membrane and subsequently incubating with a solution that contains X-gal will allow the selection of true positives.

7.2. Duration

3–4 days

3.1 Pick large colonies from the 25 15-cm yeast plates. Using a sterile micropipettor tip, place each colony into a separate microcentrifuge tube containing 100-μl sterile water.

3.2 Pick one or two colonies from one of the titration plates as negative controls for the second selection.

3.3 If possible, pick one colony that will serve as a positive control (a known interaction in the same strain of yeast, but could be a different bait and prey pair).

3.4 Vortex to mix and obtain a homogeneous liquid yeast mixture.

3.5 Spot 3 μl of each colony onto two 10-cm plates: SC-Leu-Trp-His + 3AT and SC-Leu-Trp-His-Adenine + 3AT.

3.6 Allow the plates to dry and then incubate them inverted for 48 h at 30 °C.

3.7 Examine the plates for colonies that are growing on both. These are likely to be true positive interactors based on selection by both histidine and adenine.

3.8 Monitor both the plates for another day or two, making sure that the colonies growing on SC-Leu-Trp-His-Adenine + 3AT plates do not start to turn red. Yeast that are deficient for adenine exhibit a red color. If they are deficient for adenine, then they are unable to activate expression of their adenine biosynthesis gene, indicating that they are not true positives.

See Fig. 3.4 for the flowchart of Step 3.

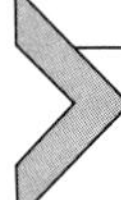

8. STEP 4 PLASMID RESCUE FROM YEAST

8.1. Overview

The pPC86 plasmid from presumed positive clones must be isolated to determine the identity of each interactor. The following is a protocol for isolating plasmid DNA from yeast and preparing it for electroporation.

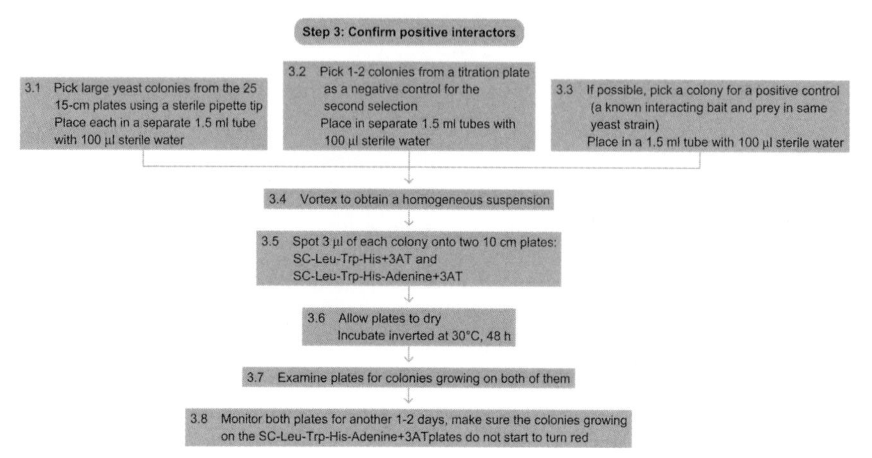

Figure 3.4 Flowchart of Step 3.

8.2. Duration

Inoculate overnight cultures + 1–2 h the next day

4.1 Prepare 2-ml overnight cultures for each positive clone in SC-Leu-Trp-His-Adenine + 3AT liquid media at 30 °C.

4.2 Collect cells by centrifuging them at 3000 rpm for 10 min. Pour off the supernatant and briefly vortex to resuspend cells in residual liquid. Transfer the cells to microcentrifuge tubes with locking caps.

4.3 Add 0.2 ml of buffer A, 0.2-ml phenol:chloroform:isoamyl alcohol (25:24:1), and about 0.2 ml of 425–600-μm glass beads.

4.4 Vortex for 2 min.

4.5 Centrifuge for 5 min at room temperature at maximum speed.

4.6 Remove ~150 μl of the aqueous top layer to a fresh 1.5-ml microcentrifuge tube.

4.7 To 150 μl of isolated plasmid DNA, add 300-μl ice-cold ethanol and incubate on ice for 15–30 min to precipitate the DNA.

4.8 Spin at 14 000 rpm for 15 min at 4 °C.

4.9 Decant the supernatant and wash the DNA pellet twice in 500-μl ice-cold 70% ethanol. Centrifuge for 5 min at room temperature between the two washes.

4.10 After the second wash, remove as much liquid as possible and briefly air-dry the DNA (5–15 min depending on how wet the DNA is).

4.11 Resuspend the DNA in 50-μl sterile water and store at −20 °C or proceed directly to electroporation of *E. coli* with the yeast DNA.

See Fig. 3.5 for the flowchart of Step 4.

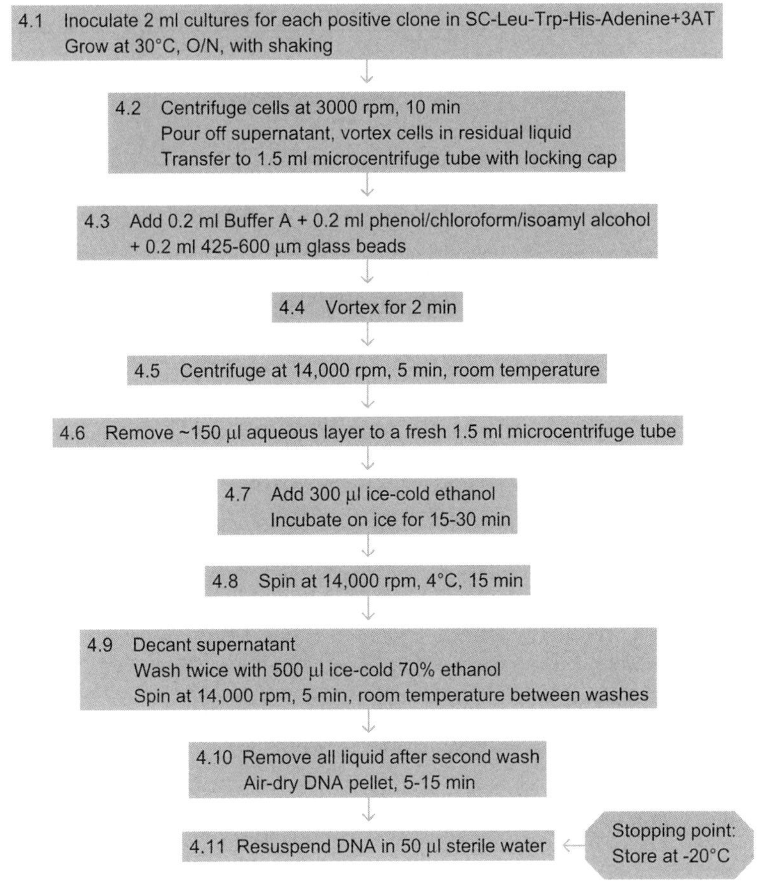

Figure 3.5 Flowchart of Step 4.

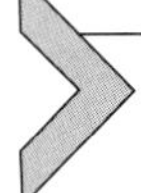

9. STEP 5 ELECTROPORATION OF *E. COLI* WITH YEAST DNA AND IDENTIFICATION OF POSITIVE INTERACTORS

9.1. Overview

DNA isolated from yeast is of poor quality and is not suitable for obtaining sequencing results. Therefore, the plasmid DNA from yeast must be electroporated into *E. coli* (see Transformation of *E. coli* via electroporation) and then isolated (see Isolation of plasmid DNA from bacteria) and sent for sequencing.

9.2. Duration

3–4 days

5.1 Chill cuvettes for electroporation on ice.

5.2 Thaw electrocompetent DH5α on ice.

5.3 Mix 1 μl of DNA and 12 μl of bacteria and transfer each mixture to a labeled cuvette.

5.4 Follow instructions for electroporation of *E. coli*.

5.5 Plate the electroporated *E. coli* on LB plates containing ampicillin to select for those carrying pPC86. Grow inverted overnight at 37 °C.

5.6 Pick two to three colonies from each transformation plate and inoculate 2–5-ml cultures of LB + ampicillin with them using a flamed inoculation loop and sterile technique. Grow cultures overnight at 37 °C, with shaking (250 rpm).

5.7 The next day, isolate the plasmid DNA using a plasmid DNA miniprep kit.

5.8 Digest the plasmid DNA with the restriction enzymes used to clone the cDNA library into the pPC86 vector. Run the digested plasmid DNAs on an agarose gel to determine the sizes of the inserts (see Agarose Gel Electrophoresis). Send the unique clones for sequencing at a DNA sequencing facility, using primers designed for use with the pPC86 vector.

5.9 When sequencing results are obtained, compare the sequences of your cDNA clones with all known transcripts of the appropriate organism using the NCBI database.

See Fig. 3.6 for the flowchart of Step 5.

10. STEP 6 BACK-TRANSFORMATION OF YEAST AND FURTHER CONFIRMATION OF INTERACTIONS

10.1. Overview

One way to further confirm that interactors are true positives is to transform yeast with the bait and identified prey of interest once again. After this method has been used to provide evidence for the existence of the interaction, biochemical and cell biological methods appropriate for the particular bait and prey of interest should be employed.

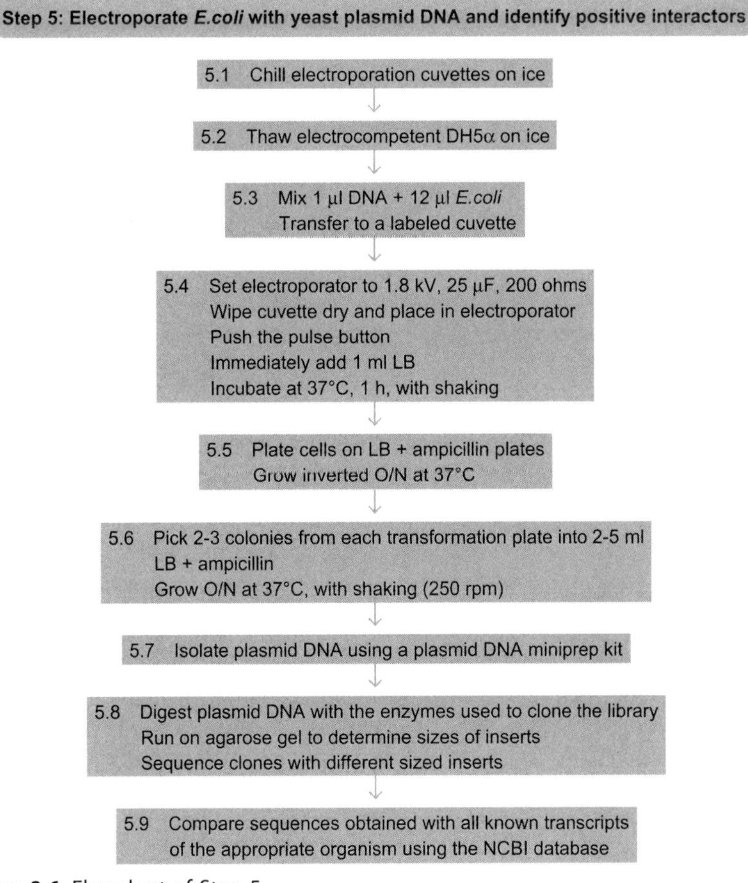

Figure 3.6 Flowchart of Step 5.

10.2. Duration

variable

6.1 Following instructions for small-scale yeast transformation, transform yeast with pDBLeu (vector alone) and with pDBLeu-X.

6.2 Next, transform these 'strains' of yeast with individual interacting clones of interest.

6.3 If they are true positive interactors, yeast transformed with a pPC86-Y and pDBLeu alone should not be able to grow on SC-Leu-Trp-His + 3AT plates, but should grow on SC-Leu-Trp. Yeast transformed with both fusion proteins should grow on both types of plates if the 'Y' cDNA encodes a true interactor.

Step 6: Back-transform yeast and further confirm positive interactors

6.1 Transform yeast with pDBLeu and pDBLeu-X (small-scale transformation as in step 1)

6.2 Transform these yeast "strains" with individual interacting clones of interest
Plate on SC-Leu-Trp and SC-Leu-Trp-His+3AT plates
Incubate at 30°C, 2-3 days

6.3 True positive interactors (yeast transformed with pDBLeu-X + pPC86-Y) will grow on both types of plate
pPC86-Y + pDBLeu-X should only grow on SC-Leu-Trp plates

6.4 Sub-clone positive interactors into mammalian expression vectors, transiently transfect cells and conduct co-immunoprecipitation experiments

Figure 3.7 Flowchart of Step 6.

6.4 To further confirm that the interactions identified in a screen are real, subclone the bait and prey cDNAs into mammalian expression vectors, and then conduct co-immunoprecipitation experiments in heterologous mammalian cells (as a guideline, you can see Transient mammalian cell Transfection with Polyethylenimine (PEI) and Analysis of Protein-Protein Interactions by Co-immunoprecipitation).

See Fig. 3.7 for the flowchart of Step 6.

REFERENCES

Related Literature

Agatep R, Kirkpatrick RD, Parchaliuk DL, Woods RA, and Gietz RD (1998) Transformation of *Saccharomyces cerevisiae* by the lithium acetate/single-stranded carrier DNA/polyethylene glycol (LiAc/ss-DNA/PEG) protocol. Technical Tips Online (http://tto.trends.com).

Johnson RC. Personal interviews. *Richard Huganir Lab*. Spring 2006.

Gietz, R. D., Triggs-Raine, B., Robbins, A., Graham, K. C., & Woods, R. A. (1997). Identification of proteins that interact with a protein of interest: Applications of the yeast two-hybrid system. *Molecular and Cellular Biochemistry*, *172*, 67–79.

Parchaliuk DL, Kirkpatrick RD, Agatep R, Simon SL, and Gietz RD (1999) Yeast two-hybrid system screening. Technical Tips Online (http://tto.trends.com).

Gietz, R. D., & Woods, R. A. (2002). Transformation of yeast by the Liac/SS carrier DNA/PEG method. *Methods in Enzymology*, *350*, 87–96.

ProQuest Two-Hybrid System Instruction Manual. Life Technologies. www.lifetech.com.

Referenced Protocols in Methods Navigator

Saccharomyces cerevisiae Growth Media.

Growth Media for *E. coli*.

Pouring Agar Plates and Streaking or Spreading to Isolate Individual Colonies.

Molecular Cloning.

Explanatory chapter: PCR -Primer design.
Transformation of Chemically Competent *E. coli*.
Chemical Transformation of Yeast.
Transformation of *E. coli* via electroporation.
Isolation of plasmid DNA from bacteria.
Agarose Gel Electrophoresis.
Transient mammalian cell Transfection with Polyethylenimine (PEI).
Analysis of Protein-Protein Interactions by Co-immunoprecipitation.

CHAPTER FOUR

UV Cross-Linking of Interacting RNA and Protein in Cultured Cells

Emi Sei, Nicholas K. Conrad[1]

Department of Microbiology, University of Texas Southwestern Medical Center, Dallas, TX, USA
[1]Corresponding Author e-mail address: nicholas.conrad@utsouthwestern.edu

Contents

Methods in Enzymology, Volume 539
ISSN 0076-6879
http://dx.doi.org/10.1016/B978-0-12-420120-0.00004-9

Abstract

RNA–protein interactions play indispensable roles in the regulation of cellular functions. Biochemical characterization of these complexes is often done by immunoprecipitation (IP) of RNA-binding proteins (RBPs) followed by identification of co-immunoprecipitated RNAs. This protocol couples ultraviolet (UV) irradiation with IP to determine whether a specific protein interacts directly with a specific RNA in living cells.

1. THEORY

RNA–protein interactions govern every aspect of RNA metabolism, including RNA processing, nuclear export, localization, stability, and translation. Experimental determination of the RBPs that associate with a specific RNA sequence is often initiated using *in vitro* or in silico analyses. Here, we describe an approach to verify that a specific RNA–protein interaction occurs in living cells.

Ribonucleoprotein particles (RNPs) can be characterized using co-IP techniques. A common assumption made when interpreting co-IP experiments is that the co-precipitation of a specific RNA with an RBP reflects an interaction that occurs in cells. Indeed, interactions observed using co-IP approaches often mirror those in live cells. However, some of these interactions result from RNA–protein reassortment in cell extracts. Cell lysis can shift the equilibria of the interactions by removing the constraints that compartmentalization and local concentrations have on complex formation *in vivo*. There have been several published examples of this phenomenon. RNA–protein complexes involved in snoRNA biogenesis (Kittur et al., 2006), RNA stability (Mili and Steitz, 2004), and viral gene expression (Sahin et al., 2010) have all been observed to reassort in cell lysates.

In order to ensure that immunoprecipitates reflect RNA–protein complexes formed *in vivo*, it is crucial to control for reassortment. To do this, co-IP experiments can be coupled with UV irradiation to covalently cross-link RNA to protein *in vivo*. The irreversible covalent bond formed upon UV cross-linking permits the use of high stringency conditions, which are necessary to ensure that RNA–protein interactions are not occurring in the lysate. Because live cells are exposed to UV, demonstration of a UV-dependent cross-link is strong evidence that a particular RBP binds a specific RNA *in vivo*. In addition, since UV irradiation only cross-links closely associated proteins and nucleic acids, positive results indicate a direct RNA–protein interaction.

One limitation of this method is that the generation of photo-induced complexes strongly depends on structural parameters such as the position of reactive amino acids or bases. Even when these parameters are met, UV cross-linking can be relatively inefficient. In addition, the cross-links are not reversible, which sometimes complicates downstream analysis. For example, protein adducts that remain linked to the RNA will decrease the efficiency of reverse transcriptase by blocking its elongation. Even so, it is clear that reverse transcriptase can extend through the cross-linked bases, so RT-PCR techniques can be compatible with UV cross-linking (Ule et al., 2005). Alternative approaches to study RNA–protein interactions formed *in vivo* include RNA immunoprecipitation (RIP) and 'cell-mixing' protocols (Niranjanakumari et al., 2002; Conrad, 2008). RIP is a formaldehyde-based cross-linking protocol that does not require direct RNA–protein interactions. Cell-mixing experiments can determine whether RNPs reassort during cell lysis. If there is no reassortment of complexes postlysis, it is not necessary to cross-link. An alternative UV cross-linking protocol (CLIP) has been developed by the Darnell lab (Ule et al., 2005). The CLIP protocol is designed to identify unknown RNAs bound to a specific RBP, whereas this protocol tests whether a particular RNA interacts with a specific protein in cells.

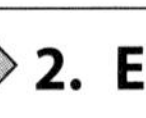

2. EQUIPMENT

Refrigerated centrifuge
Refrigerated microcentrifuge
Nutator
Pipet-aid
Aspirator
Micropipettors
UV source (Spectroline XL-1500 with 254-nm wavelength bulbs)
Heating blocks (37 and 65 °C)
Vortex mixer
15-ml conical polypropylene tubes

Pipettes
Micropipettor tips
1.5-ml microcentrifuge tubes
8″ × 8″ Pyrex® tray
Dry ice
Rubber policeman
QIAshredder spin columns (Qiagen)

3. MATERIALS

Phosphate-Buffered Saline, pH 7.2 with Mg/Ca (PBS) (Sigma)
Magnesium chloride ($MgCl_2$)
Calcium chloride ($CaCl_2$)
Sodium dodecyl sulfate (SDS)
Tris base
Hydrochloric acid (HCl)
Sodium hydroxide (NaOH)
Ethylenediaminetetraacetic acid (EDTA)
Dithiothreitol (DTT)
Sodium chloride (NaCl)
Nonidet P40 (NP40)
Sodium deoxycholate (DOC)
Sodium acetate (NaOAc)
Vanadyl-ribonucleoside complexes (VRC) (New England Biolabs)
Phenylmethylsulfonyl fluoride (PMSF)
Torula yeast RNA (cRNA) (Sigma)
β-globin RNA (or an alternative control RNA)
Anti-FLAG® M2 affinity gel (Sigma)
Proteinase K
Glycoblue (Ambion)
Phenol:chloroform:isoamyl alcohol (25:24:1)
100% ethanol

3.1. Solutions & buffers

Step 2 SDS lysis buffer

Component	Final concentration	Stock	Amount
SDS	0.5%	20%	14 μl
Tris HCl, pH 6.8	50 mM	1 M	28 μl
EDTA	1 mM	500 mM	1.12 μl

DTT	1 mM	100 mM	5.6 μl
cRNA	2.5 mg ml^{-1}	50 mg ml^{-1}	28 μl
VRC	10 mM	200 mM	28 μl

Add H_2O to 560 μl. Add 2.8 μl of 200 mM PMSF ***immediately*** before use
Note: These volumes make buffer for four samples, you will have to adjust the volumes according to your number of samples

RIPA correction buffer

Component	Final concentration	Stock	Amount
NP40	1.25%	10%	280 μl
Sodium deoxycholate	0.625%	10%	140 μl
Tris HCl, pH 8.0	62.5 mM	1 M	140 μl
EDTA	2.25 mM	500 mM	10.1 μl
NaCl	187.5 mM	5 M	84 μl
cRNA	2.5 mg ml^{-1}	50 mg ml^{-1}	112 μl
VRC	10 mM	200 mM	112 μl

Add H_2O to 2240 μl. Add 11.2 μl of 200 mM PMSF ***immediately*** before use
Note: These volumes make buffer for four samples, you will have to adjust the volumes according to your number of samples

Step 3 RIPA buffer

Component	Final concentration	Stock	Amount
NP40	1%	100%	5 ml
Sodium deoxycholate	0.5%		2.5 g
SDS	0.1%	20%	2.5 ml
NaCl	150 mM	5 M	15 ml
Tris–HCl, pH 8.0	50 mM	1 M	25 ml
EDTA	2 mM	500 mM	2 ml

Add H_2O to 500 ml

Step 4 Proteinase K solution

Component	Final concentration	Stock	Amount
Proteinase K	0.5 mg ml^{-1}	20 mg ml^{-1}	75 μl
SDS	0.5%	20%	75 μl

Tris–HCl, pH 7.5	20 mM	1 M	60 μl
EDTA	5 mM	500 mM	30 μl
Glycoblue	16.7–100 ng μl^{-1}	15 mg ml^{-1}	20 μl
cRNA	0.1 mg ml^{-1}	50 mg ml^{-1}	6 μl
β-globin RNA	6.7 pg μl^{-1}	1 ng μl^{-1}	20.1 μl

Add H_2O to 3 ml
Note: These volumes make solution for ten samples, you will have to adjust the volumes according to your number of samples

Tip	*Make all buffers fresh right before use.*
Tip	*You can add 0.3–2 μl of glycoblue to each sample.*
Tip	*We use in vitro transcribed β-globin as a loading and recovery control. However, any exogenously added RNA may be used provided it does not interfere with the detection of the RNA of interest.*

4. PROTOCOL

4.1. Duration

Preparation	**About 2 days**
Protocol	About 5–6 h

4.2. Preparation

Obtain expression constructs for the FLAG-tagged RNA binding protein and the target RNA.

A minimum of three 10 cm plates of cells will be needed for one experiment. Transfect two plates with expression constructs for both a FLAG-tagged protein and the ligand RNA. One of these will be the 'no UV' control and the other will be your 'test' sample. Transfect an additional plate with both an untagged version of your protein and the ligand RNA expression construct; this will be your 'untagged' control.

4.3. Tip

Any general transfection protocol will work; cross-linking should be performed 18–48 h posttransfection. The untagged protein control will help distinguish whether

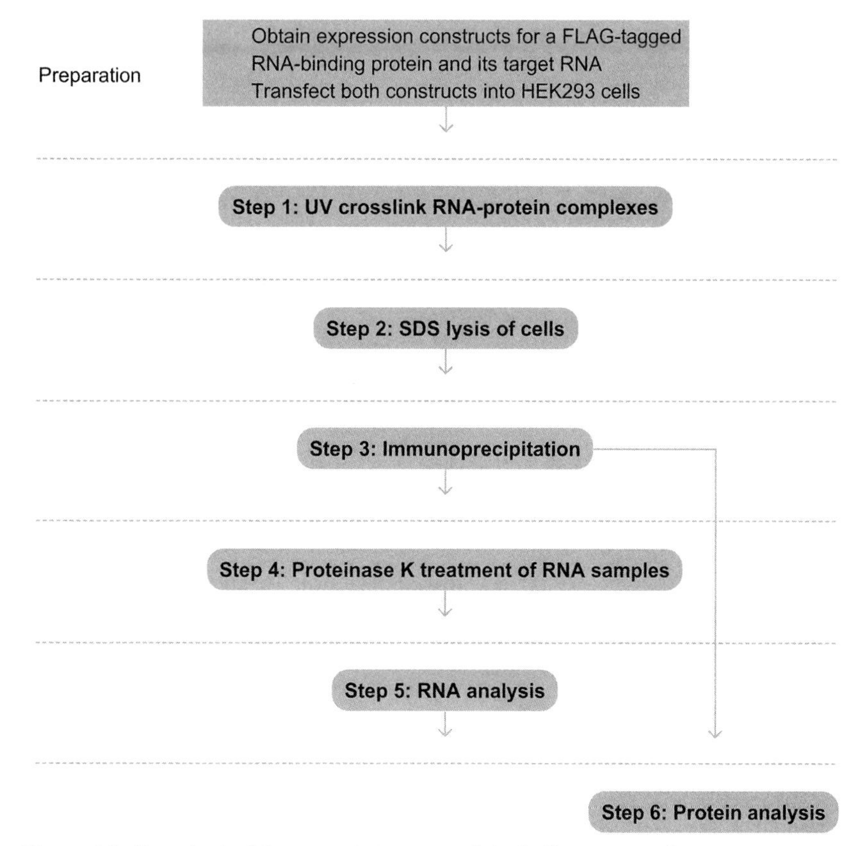

Figure 4.1 Flowchart of the complete protocol, including preparation.

any observed binding is due to nonspecific interactions with your antibody beads. If testing an endogenous RNA–protein interaction, one plate will be a 'no UV' control, the second will be your 'test' sample, and the 'untagged' sample will be replaced by a control antibody (same isotype as the one used to pull down your protein). For cells grown in suspension, use $\sim 1 \times 10^7$ *cells per condition.*

See Fig. 4.1 for the flowchart of the complete protocol.

5. STEP 1 UV CROSS-LINK RNA–PROTEIN COMPLEXES

5.1. Overview

UV cross-link the RNA–protein complexes *in vivo.*

5.2. Duration

30–45 min

1.1 Gently wash each 10 cm dish with 5 ml of ice-cold PBS (with Mg^{2+}/Ca^{2+}). Aspirate PBS.

1.2 Add 3 ml of ice-cold PBS (with Mg^{2+}/Ca^{2+}) to each dish.
1.3 Remove covers and place only the 'test' and 'untagged control' samples in the UV cross-linker, about 3 cm away from the UV source. Irradiate samples at 250 mJ cm^{-2}. See Video 4.1, http://dx.doi.org/10.1016/B978-0-12-420120-0.00004-9.
1.4 Add 7 ml of ice-cold PBS (without Mg^{2+}/Ca^{2+}) to each plate.
1.5 Scrape cells from plates using a rubber policeman and transfer suspension to 15-ml conical tubes.
1.6 Centrifuge at 700 × g at 4 °C for 3 min. Decant or aspirate supernatant.
1.7 Add 1 ml of ice-cold PBS and transfer to 1.5-ml microcentrifuge tubes.
1.8 Centrifuge at 2400 × g at 4 °C for 1 min. Remove supernatant.
1.9 Snap-freeze the pellet on dry ice. Samples can be stored at −80 °C indefinitely.

5.3. Tip

Adherent cells should be 70–100% confluent at this point. Be careful not to detach cells from plate. If using suspension cells, centrifuge at 700 × g at 4 °C for 3 min, wash cells in 10 ml of ice-cold PBS, and resuspend cell pellet as described in Step 1.2.

5.4. Tip

Make sure to level each dish during cross-linking to avoid dry areas. Keep plates on ice at all times.

5.5. Tip

This parameter might need optimization (a reasonable test range is 125–1500 mJ cm^{-2}). Dishes should be kept on ice while UV crosslinking; we place them on an 8″ × 8″ Pyrex® tray with ice.

See Fig. 4.2 for the flowchart of Step 1.

6. STEP 2 SDS LYSIS OF CELLS

6.1. Overview

Lyse the UV cross-linked cells in SDS Lysis Buffer.

6.2. Duration

1–1.5 h

2.1 Allow cell pellets to thaw on ice.
2.2 Vortex briefly (~2 s).
2.3 Add 140 μl of SDS Lysis Buffer per tube.

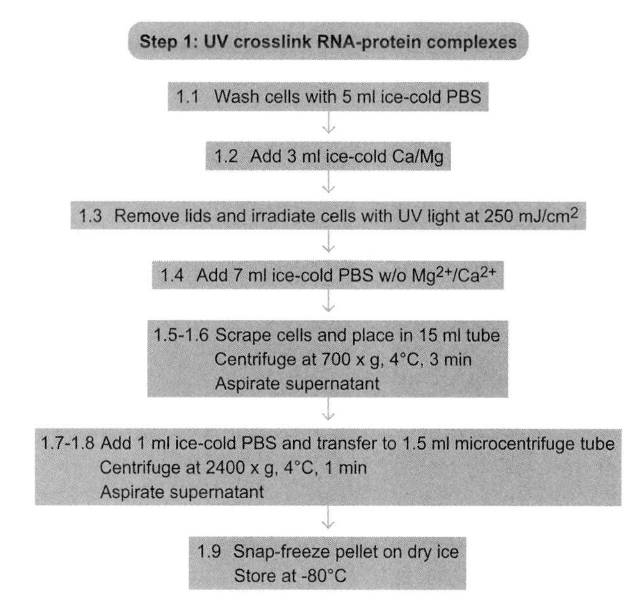

Figure 4.2 Flowchart of Step 1.

2.4 Heat samples at 65 °C for 5 min.

2.5 Chill tubes on ice for 2–3 min.

2.6 Add 560 µl ice-cold RIPA Correction Buffer.

2.7 Add solution to a QIAshredder spin column. Spin at 16000 × g at 4 °C for 1 min. Resuspend the pellet and run through the same column a second time.

2.8 Transfer flowthrough to clean 1.5-ml microcentrifuge tubes. Centrifuge at 16000 × g at 4 °C for 15 min. Repeat this step twice for a total of three spins.

6.3. Tip

Solution will be very viscous, mix as well as reasonably possible without losing sample or producing bubbles.

6.4. Tip

The solution will be very viscous, so you might have to pour the solution into the QIAshredder column. This step reduces the viscosity of the solution by shearing DNA and it also increases RNA recovery. Alternatively, sonication can be used, but this might lead to increased RNA degradation.

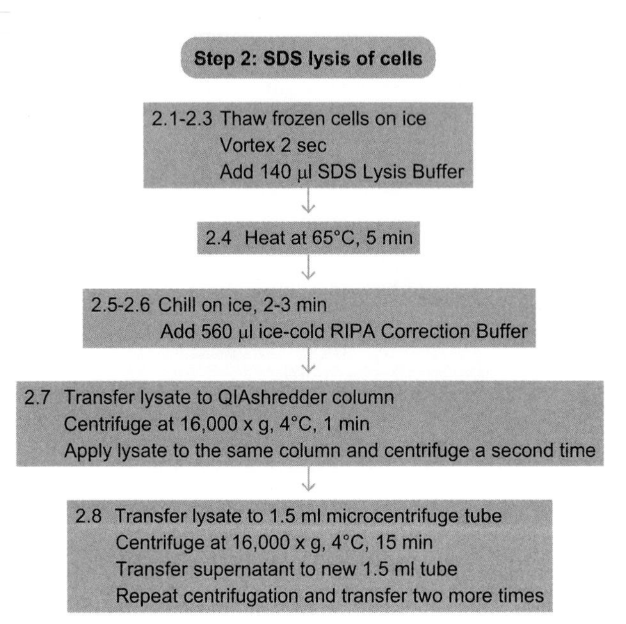

Figure 4.3 Flowchart of Step 2.

6.5. Tip

Pellet should be minimal after the third spin. In many cases, shorter centrifugation times (~5–10 min) are sufficient.

See Fig. 4.3 for the flowchart of Step 2.

7. STEP 3 IMMUNOPRECIPITATION

7.1. Overview

Immunoprecipitate the RNA-binding proteins.

7.2. Duration

~2.5 h

7.3. Preparation

Prepare the anti-FLAG M2 affinity gel. Add 40 μl of anti-FLAG M2 affinity gel slurry (~20 uL of bead volume) to a 1.5-ml microcentrifuge tube for each sample. To wash the beads, add 500 μl of RIPA buffer and centrifuge at 850 × g for 1 min. Repeat this twice for a total of three washes. After the last spin, remove all the supernatant and leave on ice until ready to continue with the immunoprecipitation.

3.1 Remove 35 μl of cell extract (from Step 2.8) for Northern (5%) and 3.5 μl for Western blot analysis (10%), and freeze aliquots on dry ice. These are your 'INPUT' samples.

3.2 Add the remaining lysate to washed anti-FLAG beads. Nutate for 2 h at 4 °C.

3.3 Centrifuge at 850 × g for 1 min at room temperature.

3.4 Remove 35 μl for Northern (5%) and 3.5 μl for Western blot analysis (10%), and freeze aliquots on dry ice. These are your 'SUPERNATANT' samples.

3.5 Discard the rest of the supernatant.

3.6 Wash the beads with 500 μl of ice-cold RIPA buffer. Centrifuge at 850 × g for 1 min at room temperature and discard the supernatant. Repeat 4 times for a total of 5 washes.

3.7 Resuspend the beads in 200 μl of RIPA buffer. Remove 10 μl of slurry and place on dry ice. This is the 'PELLET' sample for Western blot analysis.

3.8 Centrifuge the remaining slurry at 850 × g for 1 min at room temperature. Discard the supernatant. This is the 'PELLET' sample for Northern blot analysis.

7.4. Tip

If testing an endogenous interaction, the antibody can be bound to the beads using standard protocols (Harlow and Lane, 1988) and washed as described earlier.

7.5. Tip

If the 'no UV' samples are positive, it may be necessary to increase the stringency of washing by performing high salt (500 mM NaCl) and/or 1 M urea washes (Conrad, 2008).

See Fig. 4.4 for the flowchart of Step 3.

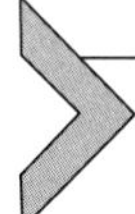

8. STEP 4 PROTEINASE K TREATMENT OF RNA SAMPLES

8.1. Overview

Digest the RNA-binding proteins with Proteinase K.

8.2. Duration

~2 h – overnight

4.1 Thaw the 'INPUT' and 'SUPERNATANT' samples that were previously frozen for Northern blot analysis.

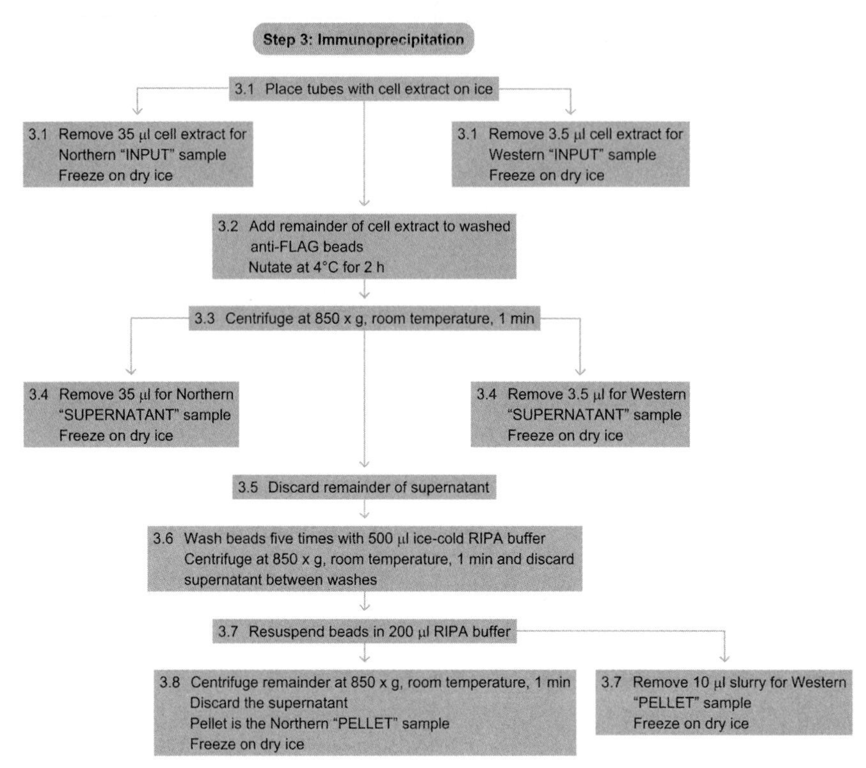

Figure 4.4 Flowchart of Step 3.

4.2 Add 300 µl of Proteinase K solution to the 'INPUT,' 'SUPERNATANT,' and 'PELLET' samples.

4.3 Incubate all samples at 37 °C for 1.5 h.

4.4 Add 30 µl of 3 M NaOAc (pH 5.2) and 350 µl of phenol/chloroform/isoamyl alcohol to each tube.

4.5 Vortex 5–10 s. Centrifuge at 16000 × g for 5 min at room temperature.

4.6 Transfer aqueous top layer to a clean 1.5-ml tube containing 900 µl of 100% ethanol. Put samples at −20 °C overnight or at −80 °C for 30 min. These samples can be stored at −20 °C indefinitely.

8.3. Tip

The proteinase K digestion can also be carried out overnight at 37 °C. Complete digestion of cross-linked proteins is crucial to ensure recovery of RNA during phenol/chloroform/isoamyl alcohol extraction.

See Fig. 4.5 for the flowchart of Step 4.

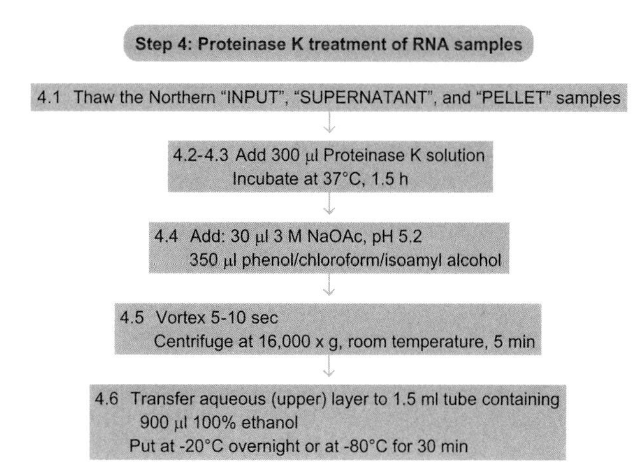

Figure 4.5 Flowchart of Step 4.

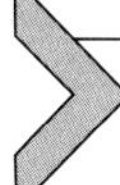

9. STEP 5 RNA ANALYSIS

9.1. Overview

Analyze the co-immunoprecipitated transcripts by Northern blot (see Northern blotting) or RNase protection assays (see Explanatory Chapter: Nuclease Protection Assays).

9.2. Duration

3–4 days

5.1 Samples from Step 4 can be analyzed by Northern blotting or RNase protection assays. About one-third to one-half of the ethanol mix is generally sufficient to observe a signal.

5.2 RNA samples should be centrifuged at 16000 × g for 15 min at room temperature, washed in 70% ethanol, and pellets should be resuspended in the buffer appropriate for the detection method of choice.

9.3. Tip

RT-PCR can be used as a detection method. However, it is important to consider that, because UV cross-links are irreversible, protein adducts will remain on the RNA after proteinase K treatment. These adducts may significantly decrease the efficiency of reverse transcriptase.

10. STEP 6 PROTEIN ANALYSIS

10.1. Overview

Analyze the immunoprecipitated proteins by Western blotting (see Western Blotting using Chemiluminescent Substrates).

10.2. Duration

1–2 days

6.1 Thaw the 'INPUT,' 'SUPERNATANT,' and 'PELLET' samples that were previously frozen for Western blot analysis.

6.2 Bring the volume of all the samples to 40 μl in standard SDS-PAGE loading buffer.

6.3 Heat 'INPUT,' 'SUPERNATANT,' and 'PELLET' samples at 100 °C for 3–5 min. Separate the proteins by SDS-PAGE (see One-dimensional SDS-Polyacrylamide Gel Electrophoresis (1D SDS-PAGE)) and perform Western blotting using standard procedures.

REFERENCES

Referenced Literature

Conrad, N. K. (2008). Co-immunoprecipitation techniques for assessing RNA–protein interactions in vivo. *Methods in Enzymology*, *449*, 317–342.

Harlow, E., & Lane, D. (1988). *Antibodies: A Laboratory Manual.* Cold Spring Harbor, NY: Cold Spring Harbor Laboratory Press.

Kittur, N., Darzacq, X., Roy, S., Singer, R. H., & Meier, U. T. (2006). Dynamic association and localization of human H/ACA RNP proteins. *RNA*, *12*(12), 2057–2062.

Mili, S., & Steitz, J. A. (2004). Evidence for reassociation of RNA-binding proteins after cell lysis: Implications for the interpretation of immunoprecipitation analysis. *RNA*, *10*(11), 1692–1694.

Niranjanakumari, S., Lasda, E., Brazas, R., & Garcia-Blanco, M. A. (2002). Reversible cross-linking combined with immunoprecipitation to study RNA–protein interactions in vivo. *Methods*, *26*(2), 182–190.

Sahin, B. B., Patel, D., & Conrad, N. K. (2010). Kaposi's sarcoma-associated herpesvirus ORF57 protein binds and protects a nuclear noncoding RNA from cellular RNA decay pathways. *PLoS Pathogens*, *6*(3), e1000799.

Ule, J., Jensen, K., Mele, A., & Darnell, R. B. (2005). CLIP: A method for identifying protein-RNA interaction sites in living cells. *Methods*, *37*(4), 376–386.

Referenced Protocols in Methods Navigator

Northern blotting.

Explanatory Chapter: Nuclease Protection Assays.

One-dimensional SDS-Polyacrylamide Gel Electrophoresis (1D SDS-PAGE).

Western Blotting using Chemiluminescent Substrates.

CHAPTER FIVE

Analysis of RNA–Protein Interactions by Cell Mixing

Sarah H. Stubbs, Nicholas K. Conrad[1]
Department of Microbiology, University of Texas Southwestern Medical Center, Dallas, TX, USA
[1]Corresponding author: e-mail address: nicholas.conrad@utsouthwestern.edu

Contents

Abstract

RNA–protein complexes are critical for almost all aspects of gene expression. Analysis of RNA–protein interactions can be complicated by the disruption of native complexes

Methods in Enzymology, Volume 539
ISSN 0076-6879
http://dx.doi.org/10.1016/B978-0-12-420120-0.00005-0

and the formation of new, reassorted complexes upon cell lysis. Before concluding that a specific RNA and protein interact *in vivo*, cell-mixing experiments can be performed to ensure that observed RNA–protein complexes are not formed after lysis of cells.

1. THEORY

Many cellular processes are regulated through RNA–protein interactions. Co-immunoprecipitation (co-IP) assays (see Co-Immunoprecipitation of proteins from yeast) are often used to characterize RNA–protein interactions, but it cannot be assumed that interactions seen through co-IP are biologically relevant. The biochemical and biophysical properties of many RNA-binding proteins promote reassortment of RNA–protein complexes in cell lysates. To determine whether an observed protein–RNA interaction forms in lysates or if it reflects an *in vivo* interaction, a cell-mixing experiment can be performed.

The methodology of a cell-mixing experiment is relatively straightforward; however, it is important to first understand the concept behind this technique. Six plates of cells will be transfected individually. Each plate will contain a different combination of any of the following plasmids expressing a tagged protein, the RNA of interest, the empty vector (for the protein), or the empty vector (for the RNA) (Fig. 5.1). In the first set of samples (plates

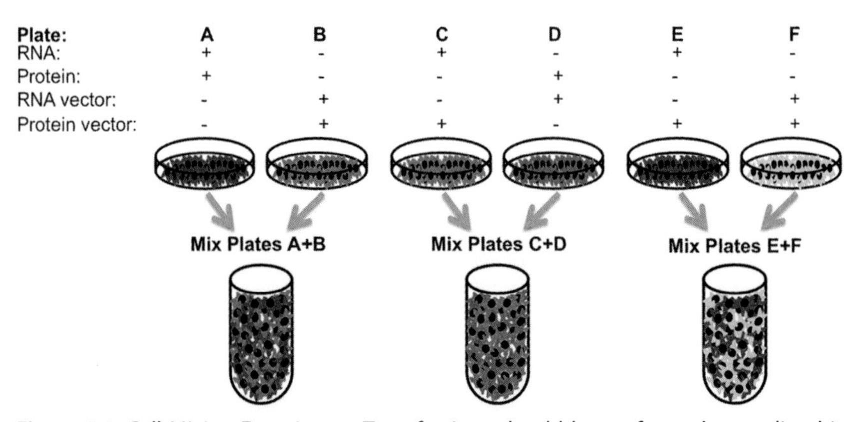

Figure 5.1 Cell-Mixing Experiment. Transfections should be performed as outlined in the figure. A+B will contain the RNA and protein of interest in the same set of cells; therefore, this should yield a positive result if the two proteins interact. As controls, C+D should be negative if no reassortment in cell extract is occurring. Additionally, E+F should be negative because no protein of interest is present. This controls for nonspecific binding of RNA to the beads. *Adapted from Conrad (2008).* (For color version of this figure, the reader is referred to the online version of this chapter.)

A and B), plate A is transfected with constructs expressing both the tagged protein and the RNA, while plate B is transfected with the corresponding empty vectors. In the next set of plates (plates C and D), cells are transfected with the construct expressing the RNA plus the empty vector for the protein (plate C) or with the construct expressing the tagged protein plus the empty vector for the RNA (plate D). After transfection, the cells are mixed, lysed, and the tagged protein is immunoprecipitated. If the interaction between the RNA and protein is formed in the lysates, the RNA will co-IP with the tagged protein, regardless of whether or not they were expressed in the same cells. Thus, similar results will be seen for A+B and C+D. If the RNA–protein complex does not form in the extract, co-immunoprecipitation will not be observed in the C+D sample. Importantly, an observed interaction in the C+D lysate should not be interpreted as evidence that the RNA–protein complex does not exist *in vivo*. Rather, this means that the particular conditions used for lysis and immunoprecipitation allow protein–RNA reassortment to occur, and thus the experiment is inconclusive as to the existence of an *in vivo* interaction. Another point to take into consideration is that the RNA can interact nonspecifically with the bead resin. To control for this, plate E is transfected with the construct expressing the RNA and plate F is transfected with the empty vectors. If nonspecific associations with the resin are occurring, the RNA will be pulled down despite the absence of the tagged protein. Variations of the transfection procedure using cell lines that naturally express the RNA or protein have also been reported (Sahin et al., 2010).

This technique has been used to demonstrate that RNA–protein interactions often reassort in cell lysate. Mili and Steitz (2004) used cell-mixing experiments to demonstrate that while the RNA binding protein, HuR, is seen to associate with *c-fos* mRNA by co-IP, a great deal of the association is due to reassortment in cell lysates. Additionally, Kittur et al. (2006) utilized the cell-mixing technique to show that a disruption of cellular compartmentalization can give false positive results. They found that NAF1, a small nucleolar ribonucleoprotein particle (snoRNP) assembly factor, is normally excluded from mature snoRNPs that are located in nucleoli and Cajal bodies. Upon cell lysis, NAF1 reassociates with snoRNPs as a result of cellular decompartmentalization. Finally, our lab used cell-mixing experiments to demonstrate that ORF57, an essential Kaposi's sarcoma-associated herpesvirus (KSHV) lytic protein, reassorts with its RNA partners in cell lysates. Importantly, the reported reassortment of RNA–protein complexes in cell extracts does not preclude the use of immunoprecipitation techniques. However, these observations underscore that co-immunoprecipitation results must be interpreted with the caveat that RNA–protein interactions in the extract do not necessarily reflect those occurring in cells.

To further support the conclusion that a particular RNA–protein interaction occurs in cells, it is necessary to validate *in vivo* RNA–protein complexes using techniques such as cross-linking methods (see UV crosslinking of interacting RNA and protein in cultured cells or PAR-CLIP (Photoactivatable Ribonucleoside-Enhanced Crosslinking and Immunoprecipitation): a step-by-step protocol to the transcriptome-wide identification of binding sites of RNA-binding proteins) or cell mixing. Cell-mixing experiments are advantageous because they are simple to perform and use a gentle cell lysis procedure, which permits the use of antibodies that are sensitive to harsh treatments necessary for cross-linking protocols. Cell-mixing experiments do not determine whether an interaction between an RNA molecule and a protein is direct or indirect, since proteins can be present in complexes. Finally, a disadvantage of cell-mixing experiments is that endogenous RNA–protein interactions generally cannot be analyzed; exogenous expression is usually required.

2. EQUIPMENT

Tissue culture incubator
Water bath (37 °C)
Refrigerated centrifuge
Refrigerated microcentrifuge
Sonicator (Branson Sonifier 450 with 4.8-mm-diameter tapered microtip)
Nutator
Vortex mixer
Aspirator
Pipet-aid
Micropipettors
15-ml polypropylene conical tubes
Pipettes
Micropipettor tips
1.5-ml microcentrifuge tubes

3. MATERIALS

Cell culture media (standard media used to grow your cell line of interest)
Transfection reagent (e.g., *Trans*IT®-293, Mirus)
Phosphate-Buffered Saline, pH 7.2 (PBS, Sigma)
10× Trypsin (Sigma)
Torula yeast RNA (cRNA, Sigma)
Phenylmethylsulfonyl fluoride (PMSF)
Protease inhibitor (Calbiochem)
RNase inhibitor (Promega)
Sodium chloride (NaCl)
Tris base
Hydrochloric acid (HCl)
Magnesium chloride ($MgCl_2$)
Triton X-100
Sodium dodecyl sulfate (SDS)
EDTA
Dithiothreitol (DTT)
Sodium acetate (NaOAc)
Phenol:chloroform:isoamyl alcohol (25:24:1)
Ethanol
Proteinase K
Glycoblue (Ambion)
Anti-FLAG M2 Beads (Sigma; alternative beads can be substituted depending on protein or tag of interest)
Dry ice

3.1. Solutions & buffers

Step 1 RSB100-T

Component	Final concentration	Stock	Amount
NaCl	100 mM	4 M	1.25 ml
Tris–HCl, pH 7.5	10 mM	1 M	500 μl
$MgCl_2$	2.5 mM	1 M	125 μl
Triton X-100	0.50%	10%	2.5 ml

Add water to 50 ml. Store at room temperature

RSB100-T-Plus

Component	Final concentration	Stock	Amount
PMSF*	1 mM	200 mM	7.5 μl
Protease Inhibitor	1 mM	100 mM	15 μl
RNase Inhibitor	0.4 U μl^{-1}	40 U μl^{-1}	15 μl
cRNA**	2.5 mg ml^{-1}	50 mg ml^{-1}	75 μl

Add RSB100-T to 1.5 ml
*Add PMSF ***immediately*** before use
**cRNA, polyU RNA, or another competitor RNA are suitable to use

Note: These volumes make enough buffer for three samples, including inputs, supernatants, and pellets being collected (nine samples total). You will have to adjust the volumes according to the number of your samples

Step 4 Proteinase K solution

Component	Final concentration	Stock	Amount
Proteinase K	0.1 mg ml^{-1}	20 mg ml^{-1}	20 μl
SDS	0.10%	20%	15 μl
Tris–HCl pH 7.5	20 mM	1 M	60 μl
EDTA	5 mM	500 mM	30 μl
Glycoblue	16.7 ng μl^{-1}	15 mg ml^{-1}	3.35 μl
cRNA	0.1 mg ml^{-1}	50 mg ml^{-1}	6 μl
Control RNA (β-globin)*	6.7 pg ml^{-1}	1 ng ml^{-1}	20.1 μl

Add water to 4 ml
*We use *in vitro* transcribed β-globin (see *in vitro* Transcription from Plasmid or PCR amplified DNA) as a loading and recovery control. However, any exogenously added RNA may be used provided it does not interfere with the detection of the RNA of interest

Note: These volumes make enough solution for nine samples; you will have to adjust the volumes according to the number of your samples

4. PROTOCOL

4.1. Duration

Preparation	About 2 days
Protocol	About 3.5–4.5 h

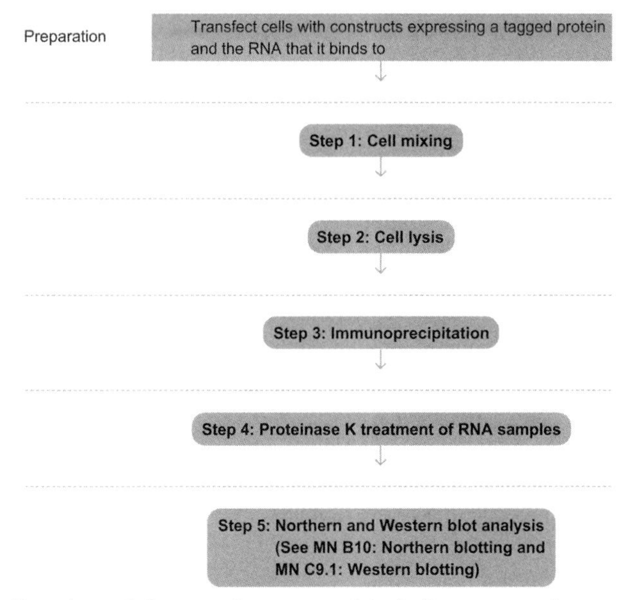

Figure 5.2 Flowchart of the complete protocol, including preparation.

4.2. Preparation

Transfect six 60 mm plates using the same amount of DNA. Refer to Figure 5.1 regarding which plasmid DNAs to use in the different transfections. Use ~6 μg of DNA per 60 mm plate. The protocol presented here uses anti-FLAG beads; other beads can be used depending on the protein being analyzed.

4.3. Tip

TransIT®-293 (Mirus) transfection reagent works well, with high transfection performance in HEK 293 cells. However, other general transfection protocols are also suitable.

See Fig. 5.2 for the flowchart of the complete protocol.

5. STEP 1 CELL MIXING

5.1. Overview

In this step, transfected cells will be harvested and mixed.

5.2. Duration

30–45 min

1.1 Prechill PBS on ice. Warm media and 1× trypsin to 37 °C. Chill centrifuge to 4 °C.
1.2 Harvest cells 18–24 h after transfection. Wash cells with 4 ml of room temperature PBS, being careful to not detach cells from the plates.
1.3 Add 1 ml of 1× trypsin to each plate. Place in 37 °C incubator for 2 min or until cells detach from the surface of the plate.
1.4 Quench trypsin by adding 4 ml of media to each plate.
1.5 Mix media plus cells from plate A with plate B in a 15-ml conical tube. Do the same for plates C+D and E+F. Refer to Figure 5.1 for a scheme of the mixing.
1.6 Centrifuge at 700 × g, 4 °C for 3 min.
1.7 During centrifugation, prepare RSB100-T-Plus for Step 2. Do not add PMSF until right before use of RSB100-T-Plus.
1.8 Aspirate media from cell pellets and wash with 5 ml of ice-cold PBS. Repellet cells at 700 × g, 4 °C for 3 min.

5.3. Tip

To ensure that cells are not detached from the plate, apply PBS on the side of the plate using a slow, steady stream of liquid.

See Fig. 5.3 for the flowchart of Step 1.

6. STEP 2 CELL LYSIS

6.1. Overview

In this step, mixed cells will be lysed (see also Lysis of mammalian and Sf9 cells) to allow cellular contents from both sets of cells to interact.

6.2. Duration

~30 min

2.1 Add PMSF to RSB100-T-Plus mixture.
2.2 Aspirate PBS, resuspend cell pellet in 400 μl of RSB100-T-Plus, and transfer cells to a 1.5-ml microcentrifuge tube.
2.3 Place the microcentrifuge tubes on ice and allow them to incubate for 5 min. During the incubation, thaw anti-FLAG beads at room temperature.
2.4 Sonicate extracts for three 5-s bursts on the lowest setting. Make sure to keep the extracts on ice for a minimum of 1 min between each round of sonication.
2.5 Centrifuge extracts at 4000 × g, 4 °C for 10 min.

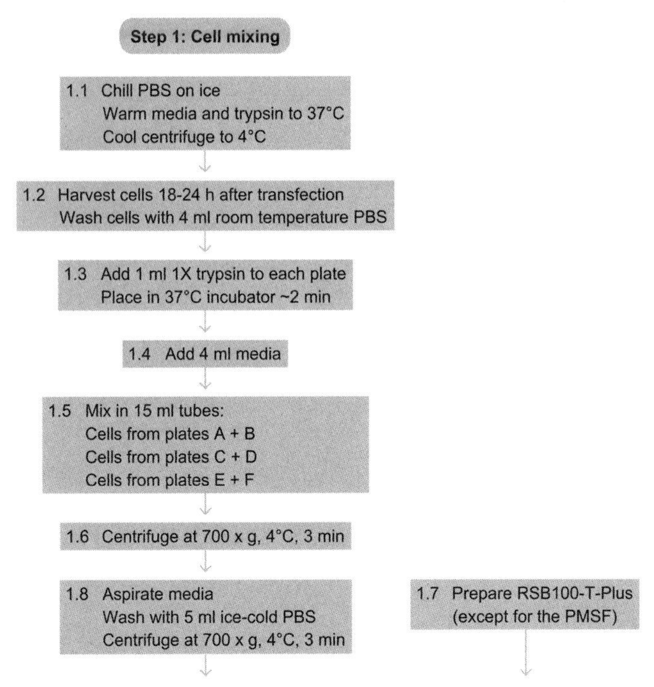

Figure 5.3 Flowchart of Step 1.

6.3. Tip

In principle, sonication can be omitted when cytoplasmic complexes are being analyzed, because RSB100-T will solubilize cytoplasmic contents.

See Fig. 5.4 for the flowchart of Step 2.

7. STEP 3 IMMUNOPRECIPITATION

7.1. Overview

Immunoprecipitation of the tagged protein bound to the RNA of interest.

7.2. Duration

~1.5 h

3.1 Wash the antibody beads during the centrifugation (Step 2.5). For each immunoprecipitation to be carried out, aliquot 40 μl of slurry (~20 μl of beads) into a 1.5-ml microcentrifuge tube. Suspend the beads in 500 μl of RSB100-T and centrifuge at 800 × g for 1 min. Wash the beads two more times using RSB100-T. After the final

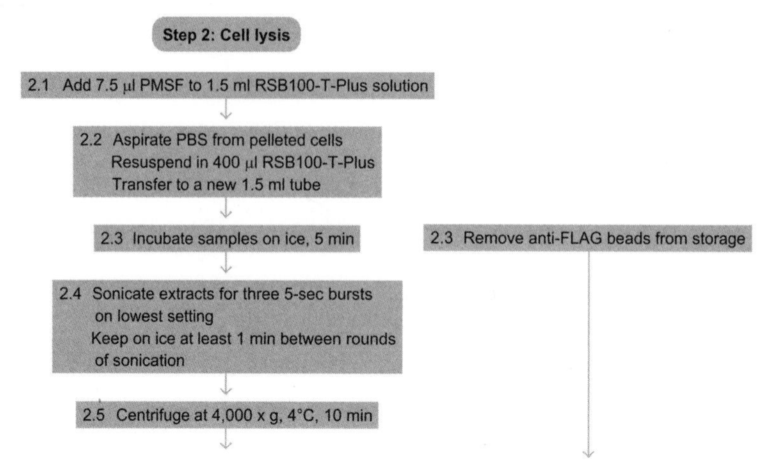

Figure 5.4 Flowchart of Step 2.

wash, resuspend the beads and aliquot into individual 1.5-ml microcentrifuge tubes for each immunoprecipitation. After dispensing the beads, leave them on ice until the samples are finished centrifuging.

3.2 Following centrifugation, transfer supernatants to new microcentrifuge tubes and centrifuge at 10000 × g, 4 °C for 10 min.

3.3 Remove INPUT samples for Northern and Western blotting. To do this, remove 20 μl of lysate (from Step 3.2) and place in a microcentrifuge tube; this is the 5% RNA INPUT. Remove 2 μl and place in a microcentrifuge tube containing SDS Loading Buffer. This is the 10% protein INPUT. Store all inputs on dry ice.

3.4 Add the remaining lysate to the washed anti-FLAG beads.

3.5 Incubate samples with rocking at 4 °C for 1 h.

3.6 Centrifuge samples at 800 × g for 1 min at room temperature.

3.7 Remove 20 and 2 μl of supernatant as above for RNA and protein SUPERNATANT samples, respectively. Discard the remaining supernatant.

3.8 Wash the beads by adding 500 μl of ice-cold RSB100-T to each sample. Spin down beads at 800 × g for 1 min. Repeat 4 times to achieve a total of 5 washes.

3.9 Remove any remaining liquid on the beads, being careful to not aspirate any beads in the pipette tip. Resuspend the beads in 200 μl of RSB100-T.

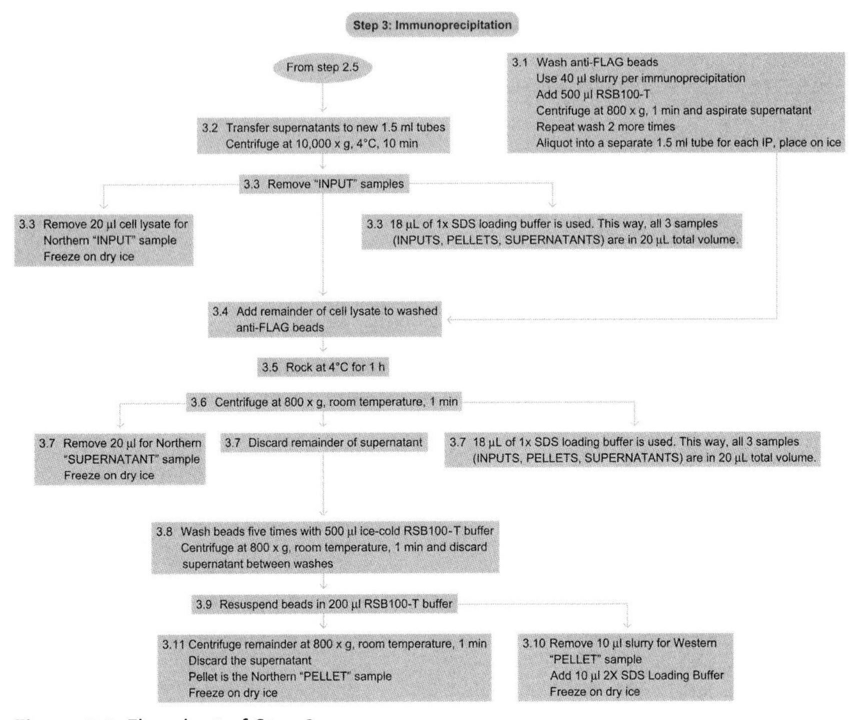

Figure 5.5 Flowchart of Step 3.

3.10 Remove 10 µl of bead slurry and place in a microcentrifuge tube. Add SDS loading buffer to each tube, this is the protein PELLET. Place samples on dry ice.

3.11 Spin down the remaining ~190 µl of beads and remove any remaining liquid, being careful not to aspirate any beads in the pipette tip. This is the RNA PELLET.

See Fig. 5.5 for the flowchart of Step 3.

8. STEP 4 PROTEINASE K TREATMENT OF RNA SAMPLES

8.1. Overview

Proteinase K digestion is performed to remove proteins prior to RNA analysis.

8.2. Duration

~1.5 h – overnight

4.1 Remove all RNA INPUT and SUPERNATANT samples from dry ice and allow them to thaw at room temperature.
4.2 Add 300 μl of Proteinase K solution to RNA INPUT, SUPERNATANT, and PELLET samples.
4.3 Incubate all samples at 37 °C for at least 30 min. If necessary, digestion can be left overnight.
4.4 Add 30 μl of 3 M NaOAc, pH 5.2, to each tube.
4.5 Add 350 μl of phenol:chloroform:isoamyl alcohol to each tube and vortex 5–10 s.
4.6 Centrifuge at 16000 × g for 5 min at room temperature.
4.7 During centrifugation, prepare microcentrifuge tubes that contain 900 μl of 100% ethanol. Transfer the top aqueous layer to tubes containing the ethanol. Before proceeding to Step 5.1, store the samples overnight at −20 °C or for 30 min at −80 °C.

8.3. Tip

Samples can be stored indefinitely in ethanol at −20 °C.

See Fig. 5.6 for the flowchart of Step 4.

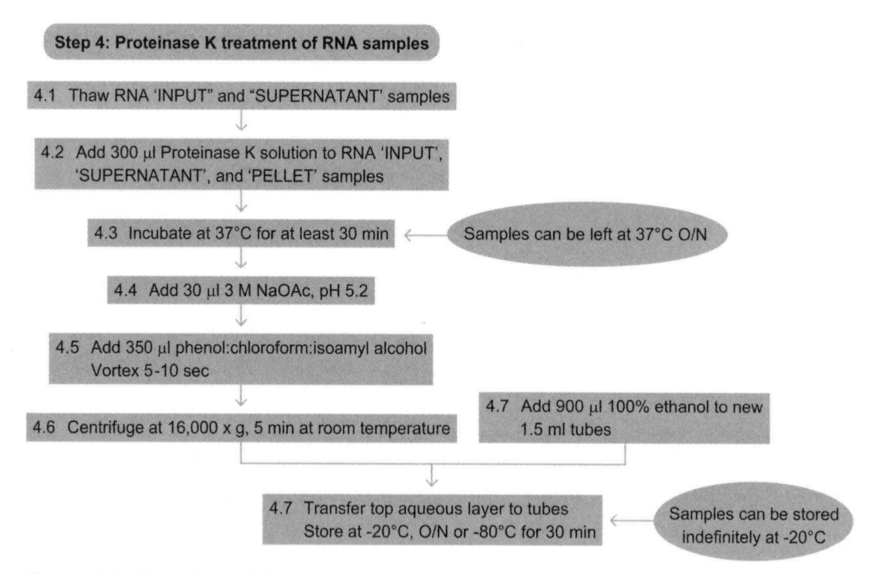

Figure 5.6 Flowchart of Step 4.

9. STEP 5 NORTHERN AND WESTERN BLOT ANALYSIS

9.1. Overview

Analyze the RNA and protein in the samples by Northern (see Northern blotting) and Western blotting (see Western Blotting using Chemiluminescent Substrates), respectively.

9.2. Duration

1–4 days

5.1 Centrifuge RNA samples at 16000 × g for 15 min, wash with 70% ethanol, and resuspend the pellet in the buffer appropriate for the detection method of choice.

5.2 RNA samples can be analyzed by Northern blotting. Quantify bands by exposing membrane to a phosphor screen, scanning on a Phosphorimager, and analyzing image using ImageQuant software, or an equivalent computer program. All samples should be normalized to the exogenously added RNA present in the Proteinase K solution. Alternatively, RNase protection assays (see Explanatory Chapter: Nuclease Protection Assays) or qRT-PCR (see Reverse-transcription PCR (RT-PCR) and Explanatory Chapter: Quantitative PCR) can be performed to quantify the RNA.

5.3 Western Blot analysis on protein samples can be performed following standard protocols. To elute proteins bound to beads, boil all samples in SDS loading buffer for 3 min.

See Fig. 5.7 for the flowchart of Step 5.

Step 5: Northern and Western blot analysis

5.1 Centrifuge RNA samples at 16,000 x g, 15 min
Wash with 70% ethanol
Resuspend in water

5.2 Analyze samples by Northern blotting
(See MN B10: Northern blotting)

5.3 Boil protein samples in SDS Loading Buffer for 3 min
Analyze samples by Western blotting
(See MN C9.1: Western blotting)

Figure 5.7 Flowchart of Step 5.

REFERENCES

Referenced Literature

Conrad, N. K. (2008). Co-immunoprecipitation techniques for assessing RNA-protein interactions in vivo. *Methods in Enzymology, 449*, 327–331.

Kittur, N., Darzacq, X., Roy, S., Singer, R., & Meier, T. (2006). Dynamic association and localization of human H/ACA RNP proteins. *RNA, 12*, 2057–2062.

Mili, S., & Steitz, J. (2004). Evidence for reassociation of RNA-binding proteins after cell lysis: Implications for the interpretation of immunoprecipitation analyses. *RNA, 10*, 1692–1694.

Sahin, B., Patel, D., & Conrad, N. (2010). Kaposi's sarcoma-associated herpesvirus ORF57 protein binds and protects a nuclear noncoding RNA from cellular RNA decay pathways. *PLoS Pathogens, 6*, e1000799.

SOURCE REFERENCES

Conrad, N. K. (2008). Co-immunoprecipitation techniques for assessing RNA-protein interactions in vivo. *Methods in Enzymology, 449*, 327–331.

Referenced Protocols in Methods Navigator

Co-Immunoprecipitation of proteins from yeast.

UV crosslinking of interacting RNA and protein in cultured cells.

PAR-CLIP (Photoactivatable Ribonucleoside-Enhanced Crosslinking and Immunoprecipitation): a step-by-step protocol to the transcriptome-wide identification of binding sites of RNA-binding proteins.

In Vitro Transcription from Plasmid or PCR amplified DNA.

Lysis of mammalian and Sf9 cells.

Northern blotting.

Western Blotting using Chemiluminescent Substrates.

Explanatory Chapter: Nuclease Protection Assays.

Reverse-transcription PCR (RT-PCR).

Explanatory Chapter: Quantitative PCR.

CHAPTER SIX

General Protein–Protein Cross-Linking

Alice Alegria-Schaffer[1]
Thermo Fisher Scientific, Rockford, IL, USA
[1]Corresponding author: e-mail address: alice.schaffer@thermofisher.com

Contents

Abstract

This protocol describes a general protein-to-protein cross-linking procedure using the water-soluble amine-reactive homobifunctional BS^3 (bis[sulfosuccinimidyl] suberate); however, the protocol can be easily adapted using other cross-linkers of similar properties. BS^3 is composed of two sulfo-NHS ester groups and an 11.4 Å linker. Sulfo-NHS ester groups react with primary amines in slightly alkaline conditions (pH 7.2–8.5) and yield stable amide bonds. The reaction releases *N*-hydroxysuccinimide (see an application of NHS esters on Labeling a protein with fluorophores using NHS ester derivitization).

Methods in Enzymology, Volume 539
ISSN 0076-6879
http://dx.doi.org/10.1016/B978-0-12-420120-0.00006-2

1. THEORY

Chemically joining two or more molecules by a covalent bond is often a necessary process for many protein research methods, including producing immunogens, creating probes for Western blotting (see Western Blotting using Chemiluminescent Substrates) and ELISA, and strategies for investigating protein structure and interactions.

Cross-linking is most often performed using near-physiologic conditions to maintain the native structure of the protein. Depending on the application, the degree of conjugation is important to consider. For example, when preparing immunogen conjugates, a high degree of conjugation is desired to increase the immunogenicity of the antigen. When conjugating to an antibody or an enzyme, maintaining the protein's biological activity is a concern and therefore, a low-to-moderate degree of conjugation is typically used. Optimal cross-linker-to-protein molar ratios for reactions must be determined empirically, although general guidelines for common applications are often included in the manufacturer's instructions.

2. EQUIPMENT

Micropipettors
Micropipettor tips
1.5-ml polypropylene tubes
Desalting columns (optional)

3. MATERIALS

Bis(sulfosuccinimidyl) suberate sodium salt (BS^3, 572.43 MW), or other amine-to-amine cross-linker
Sodium phosphate monobasic (NaH_2PO_4)
Sodium phosphate dibasic ($Na_2HPO_4 \cdot 7H_2O$)
Sodium chloride (NaCl)
Tris base
Lysine or glycine (optional)
HEPES (optional)

3.1. Solutions & buffers

Phosphate-buffered saline (PBS)*

Component	Final concentration	Stock	Amount
NaH_2PO_4	100 mM	0.2 M	14 ml
$Na_2HPO_4 \cdot 7H_2O$	0.2 M	36 ml	
Sodium chloride	150 mM		0.88 g

Adjust pH to 7.2–7.4, if needed. Add water to 100 ml
*Any non-amine-containing buffer at pH 7–9 may be used. Examples include 20-mM HEPES; 100-mM carbonate/bicarbonate; or 50-mM borate

Quenching buffer*

Component	Final concentration	Amount
Tris base	1 M	12.1 g

Add water to 100 ml. Adjust pH to 7.5 with HCl
*1-M glycine or lysine may also be used. Alternatively, remove nonreacted BS^3 by dialysis or gel filtration (e.g., desalting column, see Gel filtration chromatography (Size exclusion chromatography) of proteins)

4. PROTOCOL

4.1. Preparation

Prepare the PBS or other nonamine conjugation buffer. If the protein is in an amine-containing buffer, such as Tris or glycine, perform a buffer exchange (i.e., dialysis or desalting) to effectively remove these products. Determine protein concentration (see Quantification of Protein Concentration using UV absorbance and Coomassie Dyes).

4.2. Duration

Preparation	15 min to 3 h
Protocol	45 min to 2 h

See Fig. 6.1 for the flowchart of the complete protocol.

5. STEP 1 CALCULATE THE AMOUNT OF BS^3 TO USE

5.1. Overview

Determine the amount of BS^3 needed to achieve the appropriate molar excess.

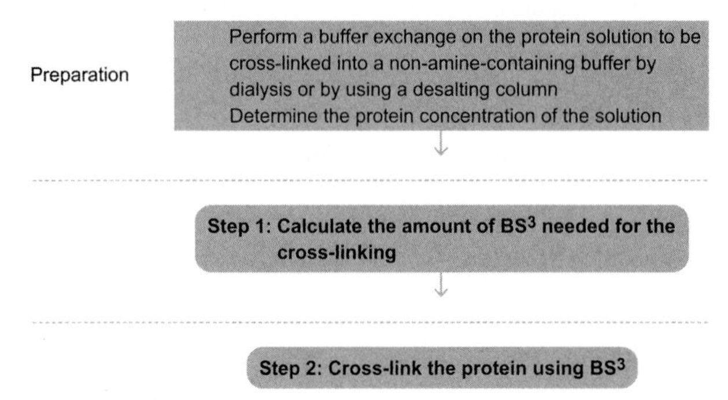

Figure 6.1 Flowchart of the complete protocol, including preparation.

5.2. Duration

1–5 min

1.1. Decide on what molar excess of BS^3 to use in the cross-linking reaction. If the protein concentration is >5 mg ml^{-1}, use a tenfold molar excess of BS^3. For samples <5 mg ml^{-1}, use a 20- to 50-fold molar excess of BS^3.

1.2 Calculate the number of millimoles of BS^3 needed to add to the reaction to give a 20-fold molar excess of BS^3 (Fig. 6.2).

1.3 Calculate the number of microliters of 25-mM BS^3 (to be prepared in Step 2.2) needed to add to the reaction (Fig. 6.2).

5.3. Tip

Hydrolysis of the NHS ester is a competing reaction and occurs more readily in dilute protein solutions. In concentrated protein solutions, the acylation reaction is favored. Therefore, cross-linking is more efficient in solutions with high protein concentrations (e.g., 5 mg ml^{-1}).

5.4. Tip

Cross-linking proteins with biological activity (e.g., enzymes, antibodies) can result in the loss of activity upon conjugation. Activity loss may also occur when the cross-linker modifies lysine groups involved in binding substrate or antigen. Adjusting the molar ratios of reagent to the target might overcome activity loss. Alternatively, use a cross-linker that targets a different functional group.

Number of millimoles of BS^3 needed to add to a 1 ml solution of 2 mg/ml IgG (150,000 MW) for a 20-fold molar excess of BS^3:

$$1\ \text{ml IgG} \times \frac{2\ \text{mg IgG}}{1\ \text{ml IgG}} \times \frac{1\ \text{mmol IgG}}{150{,}000\ \text{mg IgG}} \times \frac{20\ \text{mmol BS}^3}{1\ \text{mmol IgG}} = 0.000266\ \text{mmol BS}^3$$

Number of microliters of 25 mM BS^3 solution needed to obtain 2.66×10^{-4} mmoles of BS^3 :

$$1\ \text{liter} = 1 \times 10^6\ \mu\text{l}$$

$$\frac{25\ \text{mmol BS}^3}{1 \times 10^6\ \mu\text{l}} = \frac{2.66 \times 10^{-4}\ \text{mmol BS}^3}{X\ \mu\text{l}}$$

$$X\ \mu\text{l} = \frac{(2.66 \times 10^{-4}\ \text{mmol BS}^3)(1 \times 10^6\ \mu\text{l})}{25\ \text{mmol BS}^3}$$

$$X\ \mu\text{l} = 10.6\ \mu\text{l}$$

Figure 6.2 An example of calculations to determine how much of a 25-mM solution of BS^3 is to be added to 1 ml of a 2-mg ml^{-1} solution of IgG to give a 20-fold molar excess of BS^3.

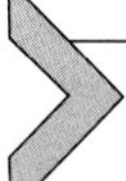

6. STEP 2 PROTEIN CROSS-LINKING

6.1. Overview

Proteins are chemically linked using BS^3.

6.2. Duration

1–2.5 h

2.1 Prepare protein in conjugation buffer (PBS or other non-amine-containing buffer).

2.2 Prepare BS^3 (572.43 MW) immediately before use. Dissolve BS^3 first in water or in a 20-mM sodium phosphate buffer. A more concentrated buffer salt might interfere with BS^3's initial solubility; however, once BS^3 is dissolved, the solution can be diluted or added to more concentrated buffer solutions without adversely affecting its solubility. Example preparations are as follows:

BS^3 concentration	Amount of water/buffer needed to add to 2-mg BS^3
12.5 mM	277 μl
25 mM	140 μl
50 mM	70 μl
100 mM	35 μl

2.3 Add the calculated amount of dissolved BS^3 to the protein sample.
2.4 Incubate the reaction mixture at room temperature for 30 min or on ice for 2 h.
2.5 Quench the reaction using by adding Quenching Buffer to a final concentration of 20–50-mM Tris. Alternatively, remove the nonreacted reagent by dialysis or desalting.
2.6 Incubate the quenching reaction at room temperature for 15 min.
2.7 Store cross-linked protein at 4 °C or −20 °C, depending on the stability of the protein, until used in downstream application(s).

6.3. Tip

BS^3 is moisture-sensitive. To avoid moisture condensation onto the product, equilibrate the vial to room temperature before opening.

6.4. Tip

Prepare BS^3 immediately before use. The NHS-ester moiety readily hydrolyzes and becomes nonreactive; therefore, do not prepare stock solutions for storage. Discard any unused reconstituted BS^3.

See Fig. 6.3 for the flowchart of Step 2.

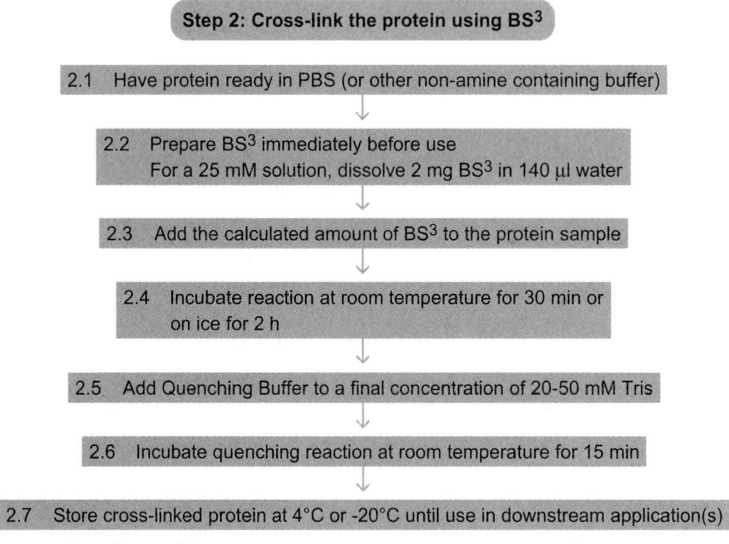

Figure 6.3 Flowchart of Step 2.

REFERENCES

Related Literature

Hermanson, G. T. (2008). In *Bioconjugate Techniques* (pp. 171–172) (2nd ed.). New York: Academic Press, and 241–243.

Pierce Biotechnology (2009) Product instructions for DSS and BS^3 Crosslinkers. Document # 0418.6.

Referenced Protocols in Methods Navigator

Labeling a protein with fluorophores using NHS ester derivitization.

Western Blotting using Chemiluminescent Substrates.

Gel filtration chromatography (Size exclusion chromatography) of proteins.

Quantification of Protein Concentration using UV absorbance and Coomassie Dyes.

CHAPTER SEVEN

Chromatin Immunoprecipitation and Multiplex Sequencing (ChIP-Seq) to Identify Global Transcription Factor Binding Sites in the Nematode *Caenorhabditis Elegans*

Cathleen M. Brdlik, Wei Niu, Michael Snyder[1]
Department of Genetics, Stanford University, Stanford, CA, USA
[1]Corresponding author: e-mail address: mpsnyder@stanford.edu

Contents

Methods in Enzymology, Volume 539
ISSN 0076-6879
http://dx.doi.org/10.1016/B978-0-12-420120-0.00007-4

Abstract

The global identification of transcription factor (TF) binding sites is a critical step in the elucidation of the functional elements of the genome. Several methods have been developed that map TF binding in human cells, yeast, and other model organisms. These methods make use of chromatin immunoprecipitation, or ChIP, and take advantage of the fact that formaldehyde fixation of living cells can be used to cross-link DNA sequences to the TFs that bind them *in vivo*. In ChIP, the cross-linked TF-DNA complexes are sheared by sonication, size fractionated, and incubated with antibody specific to the TF of interest to generate a library of TF-bound DNA sequences. ChIP-chip was the first technology developed to globally identify TF-bound DNA sequences and involves subsequent hybridization of the ChIP DNA to oligonucleotide microarrays. However, ChIP-chip proved to be costly, labor-intensive, and limited by the fixed number of probes available on the microarray chip. ChIP-Seq combines ChIP with massively parallel high-throughput sequencing (see Explanatory Chapter: Next Generation Sequencing) and has demonstrated vast improvement over ChIP-chip with respect to time and cost, signal-to-noise ratio, and resolution. In particular, multiplex sequencing can be used to achieve a higher throughput in ChIP-Seq analyses involving organisms with genomes of lower complexity than that of human (Lefrançois et al., 2009) and thereby reduce the cost and amount of time needed for each result. The multiplex ChIP-Seq method described in this section has been developed for *Caenorhabditis elegans*, but is easily adaptable for other organisms.

1. THEORY

The main objective of the Human Genome Project (HGP) was to determine the sequence of the genomes of humans, mice, *Caenorhabditis elegans*, and *Drosophila melanogaster* (Collins et al., 1998). The value of including the simpler, more compact genomes of model organisms in this project stems from their potential to provide additional insight into the conserved biological pathways of higher organisms. Though the HGP provided unprecedented scientific knowledge of these genomes through highly accurate sequence determination, a complete understanding of protein-coding regions, non-protein-coding regions, and genomic elements that regulate both temporal and spatial gene expression (ENCODE Project Consortium, 2007) remained lacking. In 2003, NHGRI launched a pilot project called ENCODE (ENCyclopedia Of DNA Elements) to identify and catalog 1% of the encoded functional elements of the human genome in a high-throughput manner. Owing to the success of the ENCODE pilot, NHGRI launched two parallel programs in 2007: a continuation of ENCODE expanded to the entire human genome and a new project called modENCODE to comprehensively annotate the functional elements of the *Caenorhabditis elegans* and *Drosophila melanogaster* genomes (modENCODE Consortium, 2009). The completion of these consortium projects is expected to give a more complete understanding of how information encoded in the genome leads to the development of a complex multicellular organism. One of the major subprojects in both of these consortia involves the global identification of the binding sites of all known transcription factors, and the technique for accomplishing this is known as ChIP-Seq.

Chromatin immunoprecipitation, or ChIP, was first developed by Varshavsky and colleagues to study protein–DNA interactions (Solomon et al., 1988). The first step of the ChIP procedure involves formaldehyde fixation of living cells to cross-link DNA sequences to the proteins that bind them *in vivo*. Formaldehyde functions by coupling primary amines that are in close proximity to one another and it is an ideal cross-linking agent for ChIP-Seq due to its short reaction time, reversible nature, and high permeability across cell membranes. Furthermore, results suggest that most of the formaldehyde-mediated protein–DNA cross-links that occur *in vivo* are due to genuine physiological interactions (Solomon and Varshavsky, 1985). After formaldehyde fixation, the resulting population of protein–DNA complexes, or chromatin, is released from cells, sonicated, and fractionated to yield an average size of 200–800 bp. Shearing serves to decrease fragment size and increase resolution. The sheared chromatin is then incubated in the presence of an antibody specific to the protein of interest to capture a library of protein-bound DNA sequences. Finally the DNA sequences are recovered following RNase A and Proteinase K digestion, reversal of cross-links, and purification. Adapters are then linked to the captured DNA, and the library is amplified for identification by ChIP-Seq or ChIP-chip.

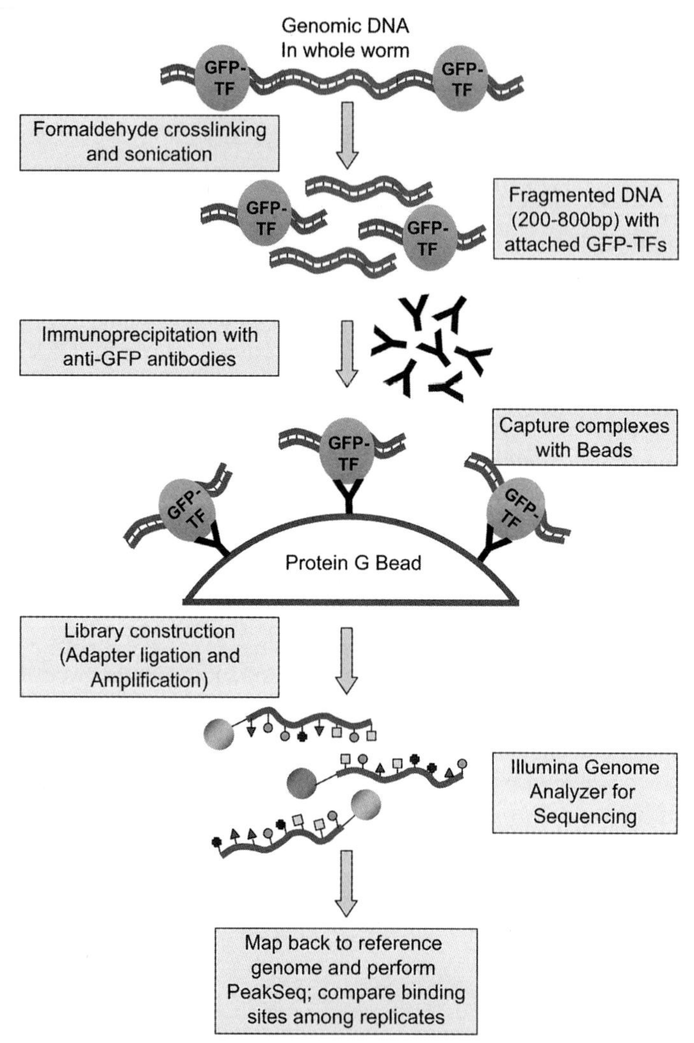

Figure 7.1 Schematic diagram of the *C. elegans* ChIP-Seq protocol. (For color version of this figure, the reader is referred to the online version of this chapter.)

ChIP-Seq (schematic in Fig. 7.1) is an extremely powerful method for mapping transcription factor binding sites that evolved from ChIP-chip, the original method developed for the global identification of transcription factor-bound DNA sequences (Ren et al., 2000; Iyer et al., 2001; Horak and Snyder, 2002). Both of these methods utilize ChIP as the first step of the procedure to obtain purified, transcription factor-bound DNA. The ChIP-chip method requires hybridization of ChIP DNA to an oligonucleotide microarray, while ChIP-Seq combines ChIP with massively parallel high-throughput sequencing (Seq) (Johnson et al., 2007; Robertson et al., 2007).

Despite being groundbreaking technology, ChIP-chip proved to be cost prohibitive because of the need for expensive oligonucleotide microarrays and hybridization experiments to be run in triplicate. In addition, microarrays have a fixed number of probes available and thus limit the number of sequences that can be identified. Although high-priced genome analyzers are necessary for ChIP-Seq, this technology is a vast improvement over ChIP-chip with respect to time and cost, signal-to-noise ratio, and resolution (Euskirchen et al., 2007). Significantly less material is needed for sequencing than for hybridization, and there is no limit to the number sequences that can be identified.

ChIP-Seq DNA samples are sequenced using the Illumina Genome Analyzer. Along with the ChIP sample, it is important to prepare an input control sample for sequencing. The input sample is a nonimmunoprecipitated control that is crucial for assessing the amount of background present. For each sample, the Illumina Genome Analyzer generates millions of short reads, approximately 30–36 nucleotides in length. These reads are mapped back to the reference genome using Illumina's ELAND software. The quality of the data (location of peaks, percantage mapping back to reference genome, number of reads, and signal-to-noise) can be visualized with a program called a 'Gene Browser' that is available from UCSC or Affymetrix. An example of data visualization in the Affymetrix IGB Browser is shown in Fig. 7.2. If the data quality is sufficient, peaks may then be called using PeakSeq (Rozowsky et al., 2009) or any of a number of other scoring algorithms (Jothi et al., 2008; Zhang et al., 2008; Tuteja et al., 2009) to indicate regions with the highest transcription factor binding. PeakSeq results are typically compared for two or more biological replicates, and the criterion for a successful data set is that 80% of the top 40% of the peaks for each replicate must overlap.

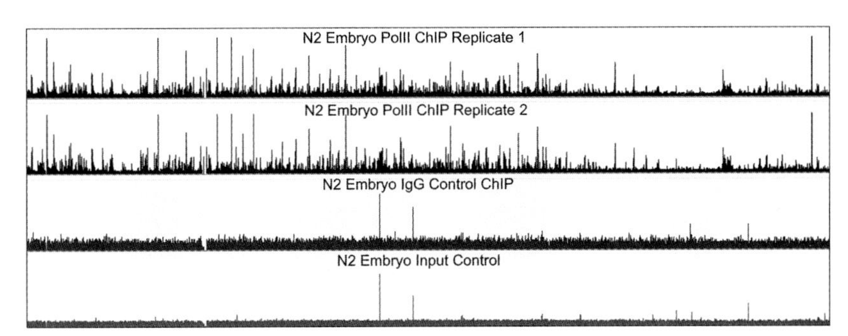

Figure 7.2 Example of ChIP-Seq data visualization using the Affymetrix IGB Browser in a region of *C. elegans* chromosome V. Experiment shown features 2 PolII ChIP replicates, an IgG ChIP control, and an input control. Note the high degree of peak overlap between PolII replicates and significant enrichment of PolII ChIP DNA in comparison to the controls. (For color version of this figure, the reader is referred to the online version of this chapter.)

The protocol described in this section was developed to perform ChIP-Seq transcription factor mapping studies with formaldehyde-fixed *Caenorhabditis elegans* preparations, but is easily adaptable to other organisms. With a completely sequenced and compact genome (100 MB, or ~1/30th that of human), a small number of cells (959) whose somatic lineage is completely defined, and powerful tools for mapping gene expression, *Caenorhabditis elegans* is an ideal model organism in which to study transcriptional regulation. The availability of tools to map gene expression is of particular interest to the modENCODE *C. elegans* Regulatory Elements Group since the strategy is to perform ChIP-Seq on stable worm lines that have been transgenically engineered to express transcription factors fused to GFP. These lines not only allow visual tracking of the GFP-fused transcription factor in the developing organism, but also permit a single antibody (anti-GFP) to be used to ChIP all of the transcription factors. Such a strategy saves enormous time and financial cost, as there is no need to screen antibodies to each transcription factor of interest. In addition, the compact size of the *C. elegans* genome requires fewer reads and accommodates the use of a multiplex short-read DNA sequencing method in which multiple samples are run in each lane of the Genome Analyzer. In this method, a unique barcode (as part of the adapter) is ligated to each library sample and four samples (each with a different barcode) are run per lane (Lefrançois et al., 2009). This method requires an additional algorithm step to parse the data into separate bins – one for each barcode – but such multiplexing capabilities have the potential to decrease cost and increase throughput by a factor of two to four.

Points to consider before performing ChIP-Seq Analysis:

1. Selection of an organism that has a completely sequenced genome and whether the genome size permits multiplex sequencing.
2. Optimization of formaldehyde fixation conditions. Under cross-linking may not keep the protein–DNA complexes intact and, in turn, not permit immunoprecipitation. Over cross-linking may interfere with DNA shearing.
3. Availability of high-quality, ChIP-grade antibodies. Having high-quality antibodies is the biggest challenge of using ChIP-Seq to identify regulatory elements. Antibodies must be tested by immunoprecipitation and immunoblotting. RNA interference followed by immunoblotting or mass spectrometry of immunoprecipitation products may also be necessary to evaluate a candidate antibody (Raha et al., 2010).
4. Optimization of the sonication protocol. Optimal sonication conditions often vary significantly among different developmental stages of

C. elegans. Therefore, it is critical to optimize sonication conditions for each kind of biological sample (cell line, developmental stage, etc.) before starting an experiment. Sonication should result in fragment sizes about 200–800 bp long, but care must be taken to keep the samples at 4 °C, thus avoiding oversonication and maintaining protein structural integrity for ChIP antibody recognition. First, the length of the sonication burst must be chosen (typically anywhere from 1 to 20 s) such that a steady 4 °C is maintained, and then the number of cycles must be varied to achieve optimal fragment size (Raha et al., 2010). An aliquot of chromatin sample is taken after each cycle interval, digested with RNase A and Proteinase K, heated to 65 °C to reverse the cross-links, purified, and then run on an agarose gel to assess fragment size. A control for determining whether the sonication conditions have left protein structure intact is to perform ChIP with RNA polymerase II antibody (Covance, cat. no. MMS-126R) and follow it with qPCR to look for enrichment of known RNA polymerase II targets.

5. Accessibility of genome mapping software and programs like PeakSeq for robust binding site scoring.

2. EQUIPMENT

Illumina Genome Analyzer
Digital sonicator with tapered microtip
-80 °C freezer
-20 °C freezer
4 °C refrigerator
Centrifuge
Vortex mixer
Spectrophotometer
Tube rotator
Nutator
Heat blocks (at 37, 55, and 65 °C)
Thermal cycler
Agarose gel electrophoresis system
Gel documentation system
NanoDrop spectrophotometer
Balance
Magnetic stir plate
Aspiration system

Micropipettors
Micropipettor tips
1.5-ml microcentrifuge tubes

3. MATERIALS

Barcoded adapters (see the oligonucleotide sequences below)
TAE 2% agarose gels (see Agarose Gel Electrophoresis for preparation)
HEPES
Potassium hydroxide (KOH)
Sodium hydroxide (NaOH)
Acetic acid
Na_2EDTA
Triton X-100
Sodium deoxycholate
Sodium chloride (NaCl)
Lithium chloride (LiCl)
NP40
Tris base
Sodium dodecyl sulfate (SDS)
Ethanol
Hydrochloric acid (HCl)
Phenylmethylsulfonyl fluoride (PMSF)
Dithiothreitol (DTT)
Bradford Reagent
N-Lauroylsarcosine sodium salt solution (20%)
dATP
DNase/RNase-free sterile water
Roche Complete Protease Inhibitor Cocktail Tablets
Antibody for chromatin immunoprecipitation
Protein A/G Sepharose Beads
RNase A
Proteinase K
Klenow fragment ($3' \rightarrow 5'$ exo-, NEB cat. # M0212S)
Phusion® High-Fidelity DNA Polymerase
Illumina PCR Primers 1.1 and 2.1
100 bp DNA ladder
QIAquick PCR Purification Kit (QIAGEN, cat. # 28104)
MinElute PCR Purification Kit (QIAGEN, cat. # 28004)
QIAquick Gel Extraction Kit (QIAGEN, cat. # 28704)

MinElute Gel Extraction Kit (QIAGEN, cat. # 28604)
End-It™ DNA End-Repair Kit (Epicentre, cat # ER0720)
LigaFast™ Rapid DNA Ligation System (Promega, cat. # M8221)

3.1. Solutions & buffers

Step 1 FA buffer

Component	Final concentration	Stock	Amount
HEPES/KOH	50 mM	1 M	50 ml
EDTA	1 mM	500 mM	2 ml
Triton X-100	1%	100%	10 ml
Sodium deoxycholate	0.1%	10%	10 ml
NaCl	150 mM	5 M	30 ml

Combine all stock solutions at the quantities listed above – omitting the Triton X-100. Bring the volume to 800 ml with ddH_2O. Then add 10 ml Triton X-100 and stir. Once the Triton X-100 is dissolved, bring the volume to 1 l with ddH_2O

FA buffer plus protease inhibitors

Component	Final concentration	Stock	Amount
FA Buffer			25 ml
PMSF	500 μM	100 mM	125 μl
DTT	1 mM	1 M	25 μl
Roche complete protease inhibitors			1 tablet

Make this buffer fresh each time

Step 2 FA-1 M NaCl buffer

Component	Final concentration	Stock	Amount
HEPES/KOH	50 mM	1 M	50 ml
EDTA	1 mM	500 mM	2 ml
Triton X-100	1%	100%	10 ml
Sodium deoxycholate	0.1%	10%	10 ml
NaCl	1 M	5 M	200 ml

Combine all stock solutions at the quantities listed above – omitting the Triton X-100. Bring the volume to 800 ml with ddH_2O. Then add 10 ml Triton X-100 and stir. Once the Triton X-100 is dissolved, bring the volume to 1 l with ddH_2O

FA-500 mM NaCl buffer

Component	Final concentration	Stock	Amount
HEPES/KOH	50 mM	1 M	50 ml
EDTA	1 mM	500 mM	2 ml
Triton X-100	1%	100%	10 ml
Sodium deoxycholate	0.1%	10%	10 ml
NaCl	500 mM	5 M	100 ml

Combine all stock solutions at the quantities listed above – omitting the Triton X-100. Bring the volume to 800 ml with ddH_2O. Then add 10 ml Triton X-100 and stir. Once the Triton X-100 is dissolved, bring the volume to 1 l with ddH_2O

TEL buffer

Component	Final concentration	Stock	Amount
LiCl	250 mM	1 M	250 ml
NP40	1%	100%	10 ml
EDTA	1 mM	500 mM	2 ml
Sodium deoxycholate	1%	10%	100 ml
Tris-HCl	10 mM	1 M	10 ml

Combine all stock solutions at the quantities listed above – omitting the NP40. Bring the volume to 800 ml with ddH_2O. Then add 10 ml NP40 and stir. Once the NP40 is dissolved, bring the volume to 1 l with ddH_2O

Elution buffer

Component	Final concentration	Stock	Amount
SDS	1%	20%	50 ml
TE	1×	10×	100 ml
NaCl	250 mM	5 M	50 ml

Combine all stock solutions at the quantities listed above, and bring the volume to 1 l with ddH_2O

Step 3 Annealing buffer for barcodes

Component	Final concentration	Stock	Amount
Tris-HCl, pH 7.5	10 mM	1 M	10 ml
NaCl	50 mM	5 M	10 ml
EDTA	1 mM	500 mM	2 ml

Combine all stock solutions at the quantities listed above, and bring the volume to 1 l with ddH_2O

3.2. Barcode sequences

Barcoding for multiplexing is as described by Lefrançois et al. (2009). Four barcodes are used for multiplexing: GTAT (MPLEXA1), CATT (MPLEXA6), ACGT (MPLEXA8), and TGCT (MPLEXA9). The following four pairs of oligos were synthesized at a 0.05-µmol scale with HPLC purification. Reverse primers (designated 'R') have a 5′-phosphate modification

MPLEXA1F	ACACTCTTTCCCTACACGACGCTCTTCCGATCTGTAT
MPLEXA1R	[Phos]TACAGATCGGAAGAGCTCGTATGCCGTCTTCTG CTTG
MPLEXA6F	ACACTCTTTCCCTACACGACGCTCTTCCGATCTCATT
MPLEXA6R	[Phos]ATGAGATCGGAAGAGCTCGTATGCCGTCTTCTG CTTG
MPLEXA8F	ACACTCTTTCCCTACACGACGCTCTTCCGATCTACGT
MPLEXA8R	[Phos]CGTAGATCGGAAGAGCTCGTATGCCGTCTTCTG CTTG
MPLEXA9F	ACACTCTTTCCCTACACGACGCTCTTCCGATCTTGCT
MPLEXA9R	[Phos]GCAAGATCGGAAGAGCTCGTATGCCGTCTTCTG CTTG

3.3. Annealing protocol

Consult the information sheet that comes with the oligos so that you know exactly how many micrograms are in each tube. Resuspend each oligo in annealing buffer to 200 µM. Mix the forward and reverse oligos for each pair in equal volumes to 100 µM and denature for 5 min at 95 °C in a wet heat block. Remove the heat block to room temperature and allow it to cool slowly over 45 min to promote annealing. Once it has cooled, dilute each adapter mix by 1:38 in DNase/RNase-free sterile water to obtain a final concentration of 30 ng μl^{-1}. Aliquot into single-use tubes (~10 µl each) to avoid freeze thawing. Use 1 µl of adapter mix at 30 ng μl^{-1} for each annealing reaction.

3.4. PCR primer preparation

Contact Illumina to purchase PCR Primers 1.1 and 2.1. Dilute the primers in DNase/RNase-free sterile water to make a stock solution at 14.3 pmoles μl^{-1}.

4. PROTOCOL

4.1. Preparation

Grow and maintain *Caenorhabditis elegans* as described by Stiernagle (2006). To achieve the quantity of worms needed for ChIP-Seq, grow gravid worms in liquid S-Basal medium overnight to yield 1.5 million embryos. Plate worms on HB101-seeded peptone-enriched plates and harvest at desired developmental stage as measured by established morphological markers.

Fix worms with 2% formaldehyde according to the protocol by Ercan et al. (2007).

4.2. Duration

Preparation	About 5–8 days to grow worms, depending on the developmental stage desired
Protocol	About 3–4 days, depending on the number of samples

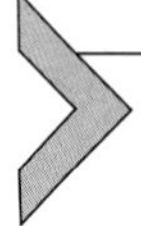

5. STEP 1 PREPARATION OF EXTRACT FROM FORMALDEHYDE-FIXED *CAENORHABDITIS ELEGANS* EMBRYOS AND LARVAE

5.1. Overview

Prepararation of formaldehyde-fixed worm pellets under conditions that will permit antibody capture of a specific transcription factor and its associated DNA sequences. Procedure begins with cell lysis and DNA shearing by sonication, followed by protein quantitation and incubation with antibody to capture the transcription factor-bound DNA.

5.2. Duration

120 min followed by a 16–20-h incubation at 4 °C (for 2 samples, if there are additional samples, add 15–20 min for each one)

1.1 Resuspend ~500 μl of packed embryos or larvae in 1.5 ml FA buffer plus protease inhibitors. Transfer to a 15-ml conical tube and bring the total volume to 3 ml with FA buffer plus protease inhibitors.

1.2 Using a digital sonicator, sonicate the samples on an ethanol/ice bath with a Branson sonifier tapered microtip at the following settings:

40% amplitude, 10 s on, and 59.9 s off for 15 cycles for embryos and 14 cycles for larvae.

1.3 Transfer the samples into microcentrifuge tubes and spin at 13 000×*g* for 15 min at 4 °C.

1.4 Transfer the supernatant to a clean tube, discarding the pellet. Determine the protein concentration by the Bradford method (see Quantification of Protein Concentration using UV absorbance and Coomassie Dyes). Continue to the next step or snap freeze the samples in liquid nitrogen and store at −80 °C.

1.5 Add the volume of clarified extract corresponding to 4.2 mg of protein to a clean microcentrifuge tube. Add 20 μl of 20% *N*-lauroylsarcosine for every 400 μl of extract and mix well.

1.6 Spin at 13 000×*g* for 5 min at 4 °C.

1.7 Transfer the supernatant to a new tube. Remove 200 μg of the material to a clean 1.5-ml microcentrifuge tube and store at −20 °C until the following day, when it will be used to prepare *input DNA*.

1.8 Split the remaining extract between two clean 1.5-ml microcentrifuge tubes to yield 2 mg each – one for the ChIP antibody and the other for the control antibody. For each immunoprecipitation, add 7.5 μg of antibody per 2 mg of extract. Rotate the tubes overnight (16–20 h) at 4 °C.

5.3. Tip

If the embryo/larvae pellet is greater than 500 μl, split the sample into multiple microcentrifuge tubes and snap-freeze the extra samples in liquid nitrogen. Store them at −80 °C.

5.4. Tip

While sonicating, make sure that the samples do not become overheated. Maintain a constant depth for the probe for all of the samples.

5.5. Tip

Make sure to remove your sample from the ethanol/ice bath as soon as sonication is completed to avoid freezing the sample.

5.6. Tip

Save the leftover FA buffer plus protease inhibitors at 4 °C for washing the beads (to follow).

See Fig. 7.3 for the flowchart of Step 1.

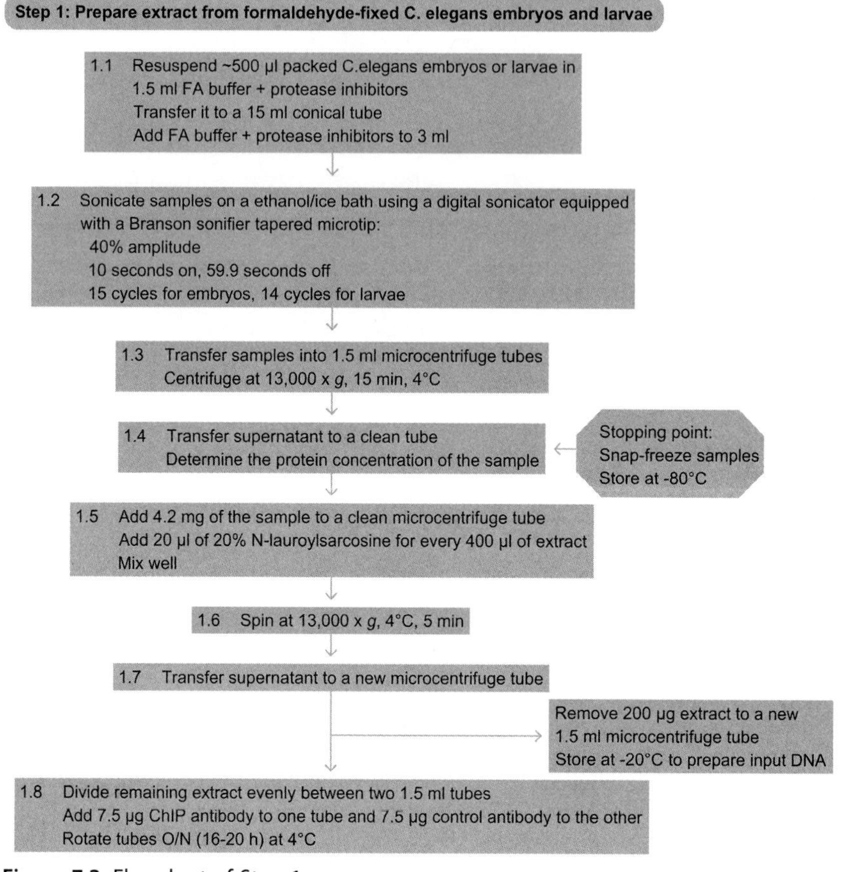

Figure 7.3 Flowchart of Step 1.

6. STEP 2 WASHES AND COLLECTION OF THE IMMUNOCOMPLEXES AND ChIP DNA PURIFICATION

6.1. Overview

Capture of the antibody–transcription factor–DNA complexes with Protein A/G sepharose beads, followed by washing, elution of the complexes, RNase A and Proteinase K digestion and reversal of the cross-links. After purification using the QIAquick PCR Purification Kit, only the DNA sequences that were formerly bound to the transcription factor of interest remain.

6.2. Duration

8 h followed by an incubation of 12–20 h (for up to 10 ChIP samples)

2.1 Aliquot 25 μl protein A or G Sepharose beads for each ChIP or control sample into separate microcentrifuge tubes and wash 4 times with 1 ml FA buffer (using the FA buffer + protease inhibitors from the day before). Spin at 2500 × *g* for 2 min to collect the beads.

2.2 Add each immunoprecipitation sample to the washed bead slurry and continue to rotate at 4 °C for 2 h.

2.3 Thaw the **input DNA** samples set aside the previous day and add 2 μl of 10 mg ml^{-1} RNase A. Digest at room temperature for 2 h.

2.4 After 2 h, add 250 μl Elution Buffer (or enough to bring the volume up to 300 μl), then add 4 μl of 10 mg ml^{-1} Proteinase K, and put the input sample at 55 °C for 4 h.

2.5 Wash the beads from the immunoprecipitations at room temperature by adding 1 ml of each of the following buffers and incubating on a nutator (or tube rotator). Collect beads by spinning for 1–2 min at 2500 × *g* and removing the supernatant:

2 times with FA buffer for 5 min each wash

1 time with FA-1 M NaCl for 5 min (After this wash, transfer thebeads to new tubes with the next wash buffer.)

1 time with FA-500 mM NaCl for 10 min

1 time with TEL buffer for 10 min

2 times with TE for 5 min each wash

2.6 To elute the immune complexes from the protein A/G Sepharose beads, add 150 μl of Elution Buffer and place the tube in a 65 °C heat block for 15 min. Briefly vortex the tube at 5 min intervals. Spin down the beads at 2500 × *g* for 2 min and transfer the supernatant to a new tube. Repeat the elution and combine the supernatants.

2.7 Add 2 μl of 10 mg ml^{-1} RNase A to each ChIP sample. Digest at room temperature for 1–2 h.

2.8 Add 2 μL of 10 mg ml^{-1} Proteinase K to each ChIP sample. Incubate for 1–2 h at 55 °C.

2.9 Transfer all input and ChIP samples to 65 °C for 12–20 h to reverse the cross-links.

2.10 Purify the DNA using the Qiaquick PCR Purification Kit. Elute with 50 μl H_2O. Run 5 μl of the input DNA on a 2% agarose TAE gel to check the extent of shearing (see Agarose Gel Electrophoresis). Most of the DNA fragments should be 200–800 bp in length.

6.3. Tip

Time the steps so as to optimize the amount of overlap (e.g., wash the ChIP bead while the input DNA undergoes Proteinase K digestion); otherwise, this step could run longer than 8 h.

6.4. Tip

Input DNA samples undergo a longer Proteinase K digestion because they have more protein – they were not purified by immunoprecipitation and washing.

6.5. Tip

Following elution of the immune complexes, remove 50 μl and resuspend in 50 μl of 2× SDS-PAGE sample buffer for analysis by SDS-PAGE and Western blotting (see One-dimensional SDS-Polyacrylamide Gel Electrophoresis (1D SDS-PAGE) and Western Blotting using Chemiluminescent Substrates).

6.6. Tip

Keep the elution volume of Qiaquick PCR purification to 50 μl so that you have enough ChIP DNA to use for qPCR later on.

See Fig. 7.4 for the flowchart of Step 2.

7. STEP 3 LIBRARY PREPARATION FOR MULTIPLEX SEQUENCING USING THE ILLUMINA GENOME ANALYZER

7.1. Overview

Due to the shear stress incurred during sonication, the ends of the DNA were damaged and must be repaired to form blunt ends. This repair step is followed by addition of an 'A' base to the 3′-end of the end-repaired DNA strands to prepare them for ligation to the barcoded adapters.

7.2. Duration

8 h (for 4 samples: 2 ChIP and 2 input DNA; this is a 2-day process for up to 20 samples: 10 ChIP and 10 input DNA)

3.1 Determine the DNA concentration of the input DNA sample and use 200 ng for each input reaction. Use 34 μl of ChIP DNA for each ChIP reaction.

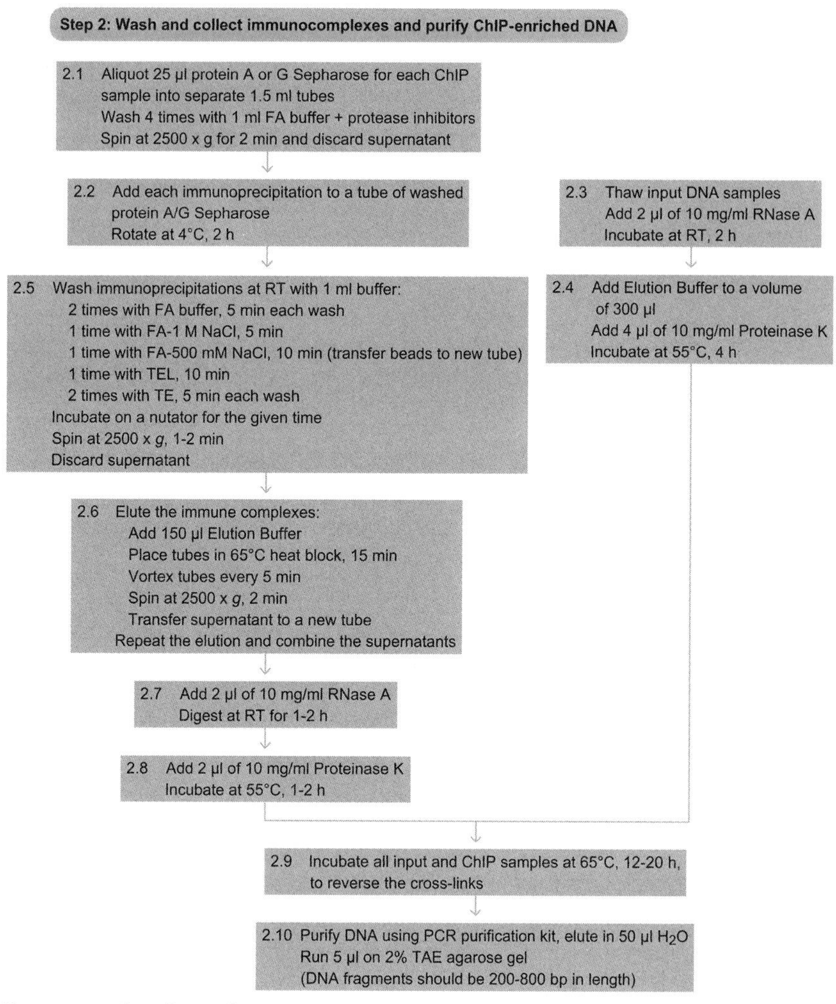

Figure 7.4 Flowchart of Step 2.

3.2 Repair the ends of the DNA molecules using End-It™ DNA End-Repair Kit. Combine the following components in a 1.5-ml microcentrifuge tube:

1–34 µl ChIP or input DNA to be end-repaired
5 µl 10× End-Repair buffer
5 µl 2.5 mM dNTP mix
5 µl 10 mM ATP
X µl sterile water (to bring reaction to 49 µl)
1 µl End-Repair enzyme mix
Incubate the reaction at room temperature for 45 min.

3.3 Purify the DNA using the QIAquick PCR Purification Kit and protocol. Elute DNA in 34 μl of EB.

3.4 Add an 'A' nucleotide to the 3′-ends using Klenow fragment (3′→5′ exo-). Combine and mix the following components in a 1.5-ml microcentrifuge tube:

34 μl DNA (from Step 3.2)
5 μl Klenow Buffer (NEB Buffer 2)
10 μl 1 mM dATP
1 μl Klenow fragment (3′→5′ exo-)
Incubate the reaction at 37 °C for 30 min.

3.5 Purify the DNA using the MinElute PCR Purification Kit according to the manufacturer's protocol. Elute DNA in 12 μl EB.

3.6 Ligate the barcoded adapters to the end-repaired DNA using the LigaFast™ Rapid DNA Ligation System. Combine and mix the following components in a 1.5-ml microcentrifuge tube (4 reactions, one for each barcoded adapter):

12 μl DNA (from Step 3.5)
15 μl 2× DNA ligase buffer
1 μl Annealed adapter oligo mix (~30 ng μl^{-1})
2 μl DNA ligase
Incubate reaction at room temperature for 15 min.

3.7 Purify the DNA using the MinElute PCR Purification Kit according to the manufacturer's protocol. Elute DNA in 18 μl EB.

3.8 Size-fractionate the ChIP and input DNA on a 2% TAE agarose gel alongside a 100 bp ladder (see Agarose Gel Electrophoresis). Take care to avoid cross-contamination by leaving an empty lane between each sample. Run the samples on the gel until there is sufficient resolution to cut a gel slice that isolates the DNA in the size range of 150–400 bp. The intended ligation product may not be visible at this stage.

3.9 Purify the DNA from the agarose slice using the QIAquick Gel Extraction Kit. Follow the instructions in the kit's manual with the following modifications: (1) Dissolve the gel slice at 37 °C – not at 50 °C as indicated. (2) Include the optional extra wash step of 500 μl Buffer QG prior to the Buffer PE wash step. (3) Elute the DNA in 25 μl EB.

3.10 Perform PCR to amplify the ChIP and input DNA libraries. Combine and mix the following components in a 0.2-ml thin-walled PCR tube:

23 μl DNA (from Step 3.9)
10 μl 5× Phusion® HF Buffer
1 μl 10 mM dNTPs
1 μl PCR primer 1.1 (14.3 pmoles μl^{-1} stock)
1 μl PCR primer 2.1 (14.3 pmoles μl^{-1} stock)
13.5 μl DNase/RNase-Free sterile water
0.5 μl Phusion® DNA Polymerase

3.11 Amplify using the following PCR protocol:
98 °C for 30 s
16 cycles of:
98 °C for 10 s
65 °C for 30 s
72 °C for 30 s
72 °C for 5 min
4 °C Hold

3.12 Purify PCR product using the MinElute PCR Purification Kit and protocol. Elute in 18 μl EB.

3.13 Size-fractionate the PCR products on a 2% TAE agarose gel alongside a 100 bp ladder (see Agarose Gel Electrophoresis). Take care to avoid cross-contamination by leaving an empty lane between each sample. Run the samples on the gel until there is sufficient resolution to cut a gel slice that isolates the DNA in the size range of 180–400 bp. The PCR product should be visible at this stage.

3.14 Purify the DNA from the agarose slice using the MinElute Gel Extraction Kit. Follow the instructions in the kit's manual with the following modifications: (1) Dissolve the gel slice at 37 °C – not at 50 °C as indicated. (2) Include the optional extra wash step of 500 μl Buffer QG prior to the Buffer PE wash step. (3) Elute the PCR product in 15–20 μl EB, maximizing the final concentration of DNA – which depends on the intensity of the PCR products on the gel.

3.15 Measure the concentration of DNA in each sample using a Nanodrop spectrophotometer. Use 2 μl of each sample for the measurement. Record and note the A260/280 ratio (should be 1.8 for pure DNA).

3.16 Mix 4 samples – 1 of each barcode – at a ratio of 1:1:1:1 (w/w/w/w) in a single tube to yield a final concentration of approximately 10 ng μl^{-1} (or higher – see Tip below). Measure the concentration one last time using the Nanodrop spectrophotometer to make sure that the mixed barcoded samples are in the correct concentration range. The sample is now ready for sequencing using the Illumina Genome Analyzer (see Explanatory Chapter: Next Generation Sequencing).

7.3. Tip

Take pictures of each agarose gel run both before *and* after *cutting the slice of interest from the gel. This documentation may aid in later troubleshooting.*

7.4. Tip

When isolating the 150–400 bp fragment (Step 3.8), be careful to cut a gel slice that does not include any DNA from an adapter–adapter band that migrates at ~120 bp.

7.5. Tip

The modifications to the QIAquick Gel Extraction Kit protocol (Step 3.9) and the MinElute Gel Extraction Kit protocol (Step 3.14) increase the extraction efficiency and cleanliness of the DNA.

7.6. Tip

If the A260/280 ratio is lower than 1.8, this suggests that there are protein impurities in your sample and hence incomplete digestion with Proteinase K. If the ratio is closer to 2.0, this suggests that there is RNA contamination in your sample (the A260/280 ratio for pure RNA is 2.0). This may reflect incomplete digestion with RNase A (for more information about absorbance ratios see Explanatory Chapter: Nucleic Acid Concentration Determination).

7.7. Tip

To calculate how much of each barcoded sample to add to the tube, start with the sample that has the lowest concentration of the four and determine how many microliters of this sample are needed to obtain 50 ng of DNA (e.g., if the concentration is 7 ng μl^{-1}, 7 μl is needed to obtain ~50 ng). Add 50 ng of each of the three remaining barcoded samples to the tube (e.g., add 5 μl of a 10 ng μl^{-1} sample, 4.5 μl of a 11 ng μl^{-1} sample, and 3 μl of a 15 ng μl^{-1} sample). To determine the final concentration, divide the total amount of DNA (in nanograms; in this case it is $50 \times 4 = 200$ ng) by the total volume (in microliters; in this case it is $7 + 5 + 4.5 + 3 = 19.5$ μl). Thus, the final concentration of the DNA is 10 ng μl^{-1}. While it may be possible to obtain usable sequence information from multiplexed libraries with a DNA concentration lower than 10 ng μl^{-1}, this is not recommended.

See Fig. 7.5 for the flowchart of Step 3.

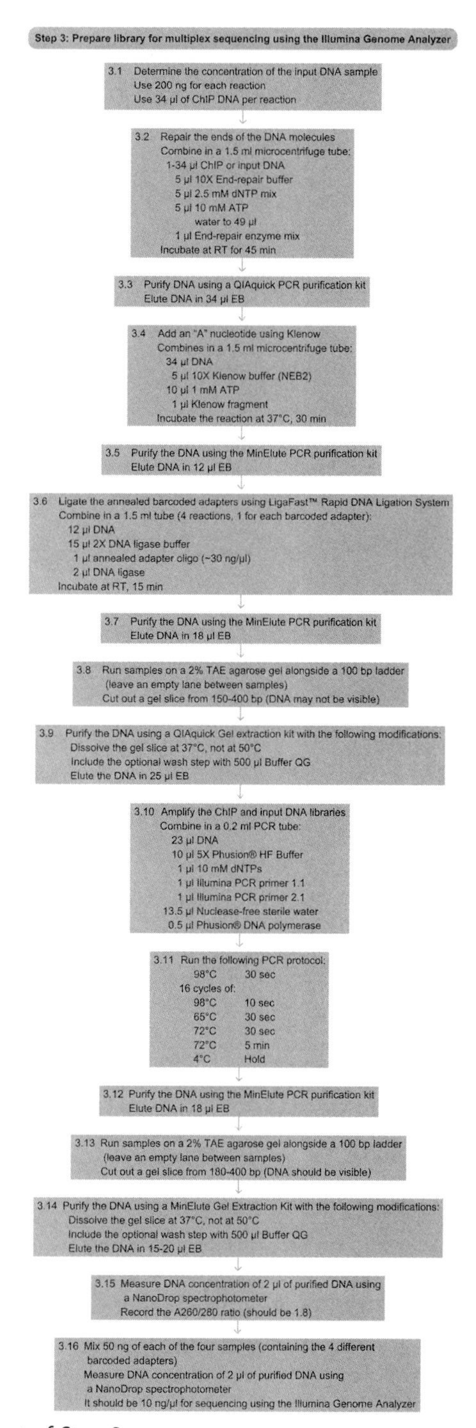

Figure 7.5 Flowchart of Step 3.

ACKNOWLEDGMENTS

The authors thank Janine Mok for reviewing the manuscript and Lixia Jiang and Beijing Wu for their efforts in generating data for the modENCODE *C. elegans* Regulatory Elements Project.

REFERENCES

Referenced Literature

Celniker, S. E., Dillon, L. A., Gerstein, M. B., Gunsalus, K. C., Henikoff, S., Karpen, G. H., et al. (2009). Unlocking the secrets of the genome. *Nature*, *459*(7249), 927–930.

Collins, F. S., Patrinos, A., Jordan, E., Chakravarti, A., Gesteland, R., & Walters, L. (1998). New Goals for the U.S. Human Genome Project: 1998–2003. *Science*, *282*(5389), 682–689.

ENCODE Project Consortium. (2007). Identification and analysis of functional elements in 1% of the human genome by the ENCODE pilot project. *Nature*, *447*(7146), 799–816.

Ercan, S., Giresi, P. G., Whittle, C. M., Zhang, X., Green, R. D., & Lieb, J. D. (2007). X chromosome repression by localization of the *C. elegans* dosage compensation machinery to sites of transcription initiation. *Nature Genetics*, *39*(3), 403–408.

Horak, C. E., & Snyder, M. (2002). ChIP-chip: A genomic approach for identifying transcription factor binding sites. *Methods in Enzymology*, *350*, 469–483.

Iyer, V. R., Horak, C. E., Scafe, C. S., Botstein, D., Snyder, M., & Brown, P. O. (2001). Genomic binding sites of the yeast cell-cycle transcription factors SBF and MBF. *Nature*, *409*(6819), 533–538.

Johnson, D. S., Mortazavi, A., Myers, R. M., & Wold, B. (2007). Genome-wide mapping of in vivo protein–DNA interactions. *Science*, *316*(5830), 1497–1502.

Jothi, R., Cuddapah, S., Barski, A., Cui, K., & Zhao, K. (2008). Genome-wide identification of in vivo protein–DNA binding sites from ChIPSeq data. *Nucleic Acids Research*, *36*, 5221–5231.

Lefrançois, P., Euskirchen, G. M., Auerbach, R. K., Rozowsky, J., Gibson, T., Yellman, C. M., et al. (2009). Efficient yeast ChIP-Seq using multiplex short-read DNA sequencing. *BMC Genomics*, *10*, 37.

Raha, D., Hong, M., & Snyder, M. (2010). ChIP-Seq: a method for global identification of regulatory elements in the genome. *Current Protocols in Molecular Biology,* Chapter 21, Unit 21.19.1–21.19.14.

Ren, B., Robert, F., Wyrick, J. J., Aparicio, O., Jennings, E. G., Simon, I., et al. (2000). Genome-wide location and function of DNA binding proteins. *Science*, *290*(5500), 2306–2309.

Robertson, G., Hirst, M., Bainbridge, M., Bilenky, M., Zhao, Y., Zeng, T., et al. (2007). Genome-wide profiles of STAT1 DNA association using chromatin immunoprecipitation and massively parallel sequencing. *Nature Methods*, *4*(8), 651–657.

Rozowsky, J., Euskirchen, G., Auerbach, R. K., Zhang, Z. D., Gibson, T., Bjornson, R., et al. (2009). PeakSeq enables systematic scoring of ChIP-seq experiments relative to controls. *Nature Biotechnology*, *27*(1), 66–75.

Solomon, M. J., Larsen, P. L., & Varshavsky, A. (1988). Mapping protein–DNA interactions in vivo with formaldehyde: Evidence that histone H4 is retained on a highly transcribed gene. *Cell*, *53*(6), 937–947.

Solomon, M. J., & Varshavsky, A. (1985). Formaldehyde-mediated DNA–protein crosslinking: A probe for in vivo chromatin structures. *Proceedings of the Natinal Academy of Sciences of the United States of America*, *82*(19), 6470–6474.

Stiernagle T (2006) Maintenance of *C. elegans*. The *C. elegans* Research Community, WormBook, doi/10.1895/wormbook.1.101.1, http://www.wormbook.org.

Tuteja, G., White, P., Schug, J., & Kaestner, K. H. (2009). Extracting transcription factor targets from ChIP-Seq data. *Nucleic Acids Research*, *37*, e113.
Zhang, Y., Liu, T., Meyer, C. A., Eeckhoute, J., Johnson, D. S., Bernstein, B. E., et al. (2008). Model-based analysis of ChIP-Seq (MACS). *Genome Biology*, *9*, R137.

Referenced Protocols in Methods Navigator

Explanatory Chapter: Next Generation Sequencing.
Agarose Gel Electrophoresis.
Quantification of Protein Concentration using UV absorbance and Coomassie Dyes.
One-dimensional SDS-Polyacrylamide Gel Electrophoresis (1D SDS-PAGE).
Western Blotting using Chemiluminescent Substrates.
Explanatory Chapter: Nucleic Acid Concentration Determination.

CHAPTER EIGHT

PAR-CLIP (Photoactivatable Ribonucleoside-Enhanced Crosslinking and Immunoprecipitation): a Step-By-Step Protocol to the Transcriptome-Wide Identification of Binding Sites of RNA-Binding Proteins

Jessica Spitzer*[,2], Markus Hafner*[,2], Markus Landthaler[3], Manuel Ascano*, Thalia Farazi*, Greg Wardle*, Jeff Nusbaum*, Mohsen Khorshid†, Lukas Burger†, Mihaela Zavolan†, Thomas Tuschl*[,1]

*Howard Hughes Medical Institute, Laboratory of RNA Molecular Biology, The Rockefeller University, New York, NY, USA
†Biozentrum der Universität Basel and Swiss Institute of Bioinformatics (SIB), Basel, Switzerland
[1]Corresponding author: e-mail address: ttuschl@rockefeller.edu
[2]These authors contributed equally to this work
[3]Present address: Berlin Institute for Medical Systems Biology, Max-Delbruck-Center for Molecular Medicine, Berlin, Germany

Contents

Methods in Enzymology, Volume 539
ISSN 0076-6879
http://dx.doi.org/10.1016/B978-0-12-420120-0.00008-6

Abstract

We recently developed a protocol for the transcriptome-wide isolation of RNA recognition elements readily applicable to any protein or ribonucleoprotein complex directly contacting RNA (including RNA helicases, polymerases, or nucleases) expressed in cell culture models either naturally or ectopically (Hafner et al., 2010).

Briefly, immunoprecipitation of the RNA-binding protein of interest is followed by isolation of the crosslinked and coimmunoprecipitated RNA. In the course of lysate preparation and immunoprecipitation, the mRNAs are partially degraded using Ribonuclease T1. The isolated crosslinked RNA fragments are converted into a cDNA library and deep-sequenced using Solexa technology (see Explanatory Chapter: Next Generation Sequencing). By introducing photoreactive nucleosides that generate characteristic sequence changes upon crosslinking (see below), our protocol allows one to separate RNA segments bound by the protein of interest from the background uncrosslinked RNAs.

1. THEORY

Posttranscriptional regulation (PTR) of messenger RNAs (mRNAs) plays important roles in diverse cellular processes (Ambros, 2004; Halbeisen et al., 2008). The fates of mRNAs are determined predominantly by their interactions with RNA-binding proteins (RBPs) and noncoding, guide-RNA-containing ribonucleoprotein complexes (RNPs). Taken together, they form mRNA-containing ribonucleoprotein complexes (mRNPs). The RBPs influence the structure and interactions of the RNAs and play critical roles in their biogenesis, stability, function, transport, and cellular localization (Moore, 2005; Keene, 2007; Glisovic et al., 2008).

Given that hundreds of RBPs and RNPs and their networks remain to be studied and evaluated in a cell-type-dependent manner, the development of powerful tools to determine their binding sites or RNA recognition elements (RREs) is critical to enhance our understanding of PTR. It offers new opportunities for understanding both gene regulation and consequences of genetic variation in transcript regions aside from the open reading frame.

Typically, a combination of genetic, biochemical, and computational approaches has been applied to identify mRNA-RBP or mRNA-RNP interactions. However, each of these methods has limitations. Microarray profiling of mRNA associated with immunopurified RBPs (RIP-ChIP) (Tenenbaum et al., 2000) is limited by incomplete enrichment of bound mRNAs and the difficulty of locating the RRE in the hundreds to thousands of nucleotide (nt) long target mRNA (Gerber et al., 2006; Landthaler et al., 2008).

Some of these problems were addressed by an *in vivo* UV 254-nm crosslinking and immunoprecipitation (CLIP) protocol (Ule et al., 2003. See also UV crosslinking of interacting RNA and protein in cultured cells) that better defines the interaction site by isolating and sequencing small RNA segments crosslinked to RBPs. However, UV 254-nm crosslinking is not efficient, and the site of crosslinking is not revealed after sequencing of the isolated RNA fragment. To separate crosslinked sites from background noise, additional control crosslinking experiments are needed, including the use of knockout cells of the protein of interest.

To overcome these limitations, we developed a new protocol referred to as PAR-CLIP (Photoactivatable-Ribonucleoside-Enhanced Crosslinking and Immunoprecipitation) (Hafner et al., 2010).

4-Thiouridine (4SU) and 6-thioguanosine (6SG) are readily incorporated into nascent RNAs by simply supplementing the media of cultured cells with the modified nucleoside (Favre et al., 1986; Bezerra and Favre, 1990). At the concentrations used in the presented protocol, neither of the tested photoreactive nucleosides showed any detectable toxic effects based on mRNA profiling or cell count. Irradiation of the cells by UV light of 365 nm leads to crosslinking of photoreactive nucleoside-labeled cellular RNAs to interacting RBPs. Using similar irradiation protocols, 4SU incorporation substantially enhances RNA recovery compared to UV 254-nm crosslinking, 6SG performs in between these two methods.

Most importantly, the sites of crosslinking can be easily identified by mapping characteristic T to C mutations (G to A in the case of 6SG, though less pronounced) in the sequenced cDNA libraries obtained from the recovered RNA initiated by the photocrosslinking itself. We presume that the structural change upon crosslinking of the modified nucleosides to aromatic amino acid side chains directs the incorporation of a noncognate deoxynucleoside during reverse transcription of crosslinked RNAs. The presence of the mutations in sequence reads, together with the observation that multiple positions within a cluster of sequence reads can be altered, facilitates the separation from clusters of unaltered background sequences typically derived from abundant cellular RNAs.

For details on the bioinformatic analyses, please refer to our recent publication (Hafner et al., 2010).

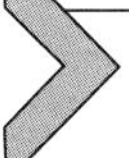

2. EQUIPMENT

Major equipment	Radioisotope laboratory
	365-nm UV-transilluminator
	Agarose gel electrophoresis equipment
	Protein electrophoresis equipment
	Equipment to cast and run 15 × 17 cm × 0.8 mm (or similar) polyacrylamide gels
	Balances (e.g., 0.1 mg–64 g and 0.1 g–4.2 kg)
	Heating block (90 °C)
	CO_2 incubator for mammalian cell culture
	D-Tube Dialyzer Midi rack (EMD Biosciences, 71511-3)
	High-speed floor centrifuge (capable of at least 13 000 × *g*)
	Magnetic rack for 1.5-ml microcentrifuge tubes and 15-ml conical tubes
	Multichannel pipettor
	pH-meter
	Phosphorimager & imaging plates (or developer and X-ray film)
	Refrigerated bench top microcentrifuge

	Rotating wheel
	PCR thermocycler
	Thermometer
	Thermomixer
	UV Stratalinker 2400 equipped with 365-nm light bulbs for crosslinking (Stratagene)
	Vortex mixer
	Water bath
	Water filter; MilliQBiocel water purification system
	X-ray exposure cassette
	HPLC with a Supelco Discovery C18 (bonded phase silica 5 μM particle, 250 × 4.6 mm) reverse phase column (Bellefonte, PA, USA)

Consumables	1.5-ml polypropylene tubes
	1.5-ml siliconized tubes (BIO PLAS Inc., 4165SL)
	15- and 50-ml conical tubes (e.g., Falcon) as well as tubes withstanding high-speed centrifugation (e.g., Sarsted, 13-ml centrifuge tube, 55.518)
	15-cm culture dishes
	10-cm tissue culture dishes
	5-μm Supor membrane syringe filter (Pall Acrodisc)
	Cell scraper (Corning)
	D-Tube Dialyzer Midi, MWCO 3.5 kDa (EMD Biosciences, 71506-3)
	NuPAGE Novex 4–12% BT Midi 1.0 gel (Invitrogen)
	pH paper (covering the range between pH 6.5 and 10)
	Plastic wrap
	Scalpels or razor blades
	Strips of 0.2-ml tubes (Thermo Scientific, AB-0264)
	Syringes (10 ml)

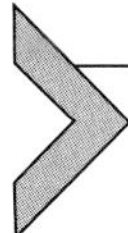

3. MATERIALS

Reagents & Chemicals	Appropriate cell culture medium and selection antibiotics
	2-Mercaptoethanol (14.3 M; Sigma, M6250)
	Acetonitrile
	Agarose, electrophoresis grade (SeaKem LE Agarose, Lonza, 50004)
	Agarose, low melting (NuSieve GTG Agarose, Lonza, 50080)
	Ammonium persulfate (APS)
	Adenosine triphosphate (ATP)
	Bacterial Alkaline Phosphatase (Worthington Biochemical, LS006344)
	Bromophenol blue
	Bovine serum albumin, acetylated (BSA, acetylated; Ambion, AM2614)
	Calcium chloride ($CaCl_2 \cdot 2H_2O$)
	Calf intestinal alkaline phosphatase (CIP)
	Chloroform
	Citric acid monohydrate
	Complete EDTA-free protease inhibitor cocktail (Roche)
	Dimethyl sulfoxide (DMSO)
	DNA ladder (25 bp)
	dNTPs: dATP, dCTP, dGTP, dTTP (0.1 M each; Fermentas, R0182)
	Dithiothreitol (DTT)
	Dynabeads Protein G (Invitrogen, 100-03D)
	EDTA disodium salt dihydrate (Sigma, E5134)
	EGTA ($C_{14}H_{20}N_2O_{10}Na_4$; Sigma, E8145-10G)
	Ethanol (100%)

Ethidium bromide
Ficoll type 400
Formamide
[γ-^{32}P]-ATP (10 mCi ml^{-1}, 6000 Ci (222 TBq) mmol^{-1}; Perkin Elmer, NEG002Z500UC)
Glycerol
Glycoblue or glycogen
Hydrochloric acid (HCl) (Fisher Scientific, A144S)
HEPES
Isoamyl alcohol
Isopropyl alcohol
Potassium chloride (KCl)
Potassium hydroxide (KOH)
Magnesium chloride ($MgCl_2 \cdot 6H_2O$)
MOPS SDS running buffer (20×; Invitrogen)
Sodium acetate (NaOAc, Fisher, S210)
Sodium phosphate dibasic ($Na_2HPO_4 \cdot 7H_2O$; Sigma, S9390-100G)
Sodium chloride (NaCl)
Sodium fluoride (NaF)
Sodium hydroxide (NaOH)
NP40 substitute (100%; Sigma [74385])
Phosphate buffered saline (PBS; 10×, commercially available)
Phenol (saturated with 0.1 M citrate buffer, pH 4.3 ± 0.2, Sigma, P4682)
Photoreactive nucleoside (Sigma; 4-thiouridine [T4509]/6-thioguanosine [858412])
Protein ladder (e.g., Biorad, 161-0374; 10–250 kDa)
Proteinase K (lyophilizate; Roche, 03115801001)
rA, rG, rC, rU, and 4SU (Sigma, T4509)
RNase T1 (Fermentas, EN0541); concentration 1000 U μl^{-1}
Sodium dodecyl sulfate (SDS; Fisher Scientific, BP166-500)

	Snake Venom Phosphodiesterase (Worthington Biochemical, LS003926)
	SuperScript III Reverse Transcriptase (Invitrogen, 18080-044); includes 5× First-strand buffer
	T4 Polynucleotide Kinase (T4 PNK; NEB, M0201)
	T4 RNA Ligase 1 (NEB, M0204L)
	T4 RNA Ligase 2, truncated (e.g., NEB, M0242L); or: Rnl2(1-249)K227Q (our plasmid for expression of the his-tagged mutant is available at www.addgene.com, plasmid 14072; however, the purified enzyme will shortly also be available from NEB)
	Taq DNA polymerase (5 U μl^{-1})
	Tris-Borate–EDTA buffer solution (TBE)
	Acetic acid – triethylamine solution 1:1 (TEAA; Sigma, 09748)
	Tetramethylethylenediamine (TEMED)
	Tris base (Fisher Scientific, BP152-1)
	Tris–HCl (Promega, H5121)
	Triton X-100
	TRIzol reagent (Invitrogen, 15596-026)
	UreaGel – SequaGel – System, National Diagnostics, EC-833
	Antibody (e.g., for FLAG-tagged RBPs: mouse monoclonal anti-FLAG M2 (Sigma, F1804))
	QIAquick gel purification kit (Qiagen)

RNA & DNA oligonucleotides	3′ adapter (DNA, except for the 5′ riboadenylate (rApp) residue): 5′ rAppTCGTATGCCGTCTTCTGCTTGT
	5′ adapter (RNA): 5′ GUUCAGAGUUCUACAGUCCGACGAUC
	3′ PCR primer (DNA): 5′ CAAGCAGAAGACGGCATACGA
	5′ PCR primer (DNA): 5′ AATGATACGGCGACCACCGACAGGTTCAGAGTTCTACAGTCCGA
	19-nt size marker (RNA): 5′ CGUACGCGGGUUUAAACGA

	24-nt size marker (RNA): 5′ CGUACGCGGAAUAGUUUAAACUGU
	33-nt size marker (RNA): 5′ CAUCUUGGUCGUACGCGGAAUAGUUUAAACUGU
	35-nt size marker (RNA): 5′CUCAUCUUGGUCGUACGCGGAAUAGUUUAA ACUGU
Reference oligoribonucleotides	CGUACGCGGAAUACUUCGA(**4SU**)U (e.g., from Thermo Scientific)
	CGUACGCGGAAUACUUCGAUU

3.1. Solutions & buffers

Step 1 1 M 4-Thiouridine stock solution

Dissolve 250-mg 4-thiouridine in 960.5-μl DMSO
(For a 1 M 6-thioguanosine solution, first dehydrate the powder supplied by Sigma in a vacuum oven at room temperature overnight. Then, dissolve 299.3 mg in 1-ml DMSO)

4-Thiouridine-containing growth medium

Component	Final concentration	Stock	Amount
4-thiouridine (in DMSO)	100 μM	1 M	100 μl
Cell culture medium			1 l

1× PBS

Component	Final concentration	Stock	Amount
PBS	1×	10×	100 ml
H_2O			900 ml

Step 2 NP40 lysis buffer

Component	Final concentration	Stock	Amount
HEPES-KOH, pH 7.5	50 mM	1 M	50 ml
KCl	150 mM	1 M	150 ml

EDTA–NaOH, pH 8.0	2 mM	0.5 M	4 ml
NaF	1 mM	0.5 M	2 ml
NP40 substitute	0.5% (v/v)	100%	5 ml
H_2O			788.5 ml
DTT (add fresh)	0.5 mM	1 M	0.5 ml
Complete EDTA-free protease inhibitor cocktail (add fresh)			1 tablet/50 ml

Step 3 Citrate-phosphate buffer, pH 5.0

Component	Amount
Citric acid monohydrate	4.7 g
$Na_2HPO_4{\cdot}7H_2O$	9.2 g
H_2O to 1 l	

Step 4 IP-wash buffer

Component	Final concentration	Stock	Amount
HEPES-KOH, pH 7.5	50 mM	1 M	50 ml
KCl	300 mM	1 M	300 ml
NP40 substitute	0.05% (v/v)	100%	0.5 ml
H_2O			649 ml
DTT (add fresh)	0.5 mM	1 M	0.5 ml
Complete EDTA-free protease inhibitor cocktail (add fresh)			1 tablet/ 50 ml

High-salt wash buffer

Component	Final concentration	Stock	Amount
HEPES-KOH, pH 7.5	50 mM	1 M	50 ml
KCl	500 mM	1 M	500 ml
NP40 substitute	0.05% (v/v)	100%	0.5 ml
H_2O			449 ml

DTT (add fresh)	0.5 mM	1 M	0.5 ml
Complete EDTA-free protease inhibitor cocktail (add fresh)			1 tablet/ 50 ml

10× Dephosphorylation buffer

Component	Final concentration	Stock	Amount
Tris–HCl, pH 7.9	50 mM	1 M	50 ml
NaCl	100 mM	3 M	33.3 ml
$MgCl_2 \cdot 6H_2O$	10 mM	1 M	10 ml
H_2O			906.2 ml
DTT (add fresh)	1 mM	1 M	0.5 ml

Step 5 Phosphatase wash buffer

Component	Final concentration	Stock	Amount
Tris–HCl, pH 7.5	50 mM	1 M	50 ml
EGTA–NaOH, pH 7.5	20 mM	0.5 M	40 ml
NP40 substitute	0.5% (v/v)	100%	5 ml
H_2O			905 ml

Polynucleotide kinase (PNK) buffer without DTT

Component	Final concentration	Stock	Amount
Tris–HCl, pH 7.5	50 mM	1 M	50 ml
NaCl	50 mM	3 M	16.7 ml
$MgCl_2 \cdot 6H_2O$	10 mM	1 M	10 ml
H_2O			923.3 ml

PNK buffer with DTT

Component	Final concentration	Stock	Amount
Tris–HCl, pH 7.5	50 mM	1 M	50 ml
NaCl	50 mM	3 M	16.7 ml
$MgCl_2 \cdot 6H_2O$	10 mM	1 M	10 ml

H_2O			918.3 ml
DTT (add fresh)	5 mM	1 M	5 ml

1× SDS PAGE loading buffer

Component	Final concentration	Stock	Amount
Tris–HCl, pH 6.8	50 mM	1 M	0.5 ml
EDTA–NaOH, pH 8.0	2 mM	0.5 M	0.04 ml
Glycerol	10% (v/v)	50%	2 ml
SDS	2% (v/v)	20%	1 ml
DTT	100 mM	1 M	1 ml
Bromophenol blue	0.1% (w/v)		10 mg
H_2O to 10 ml			

Step 6 1× MOPS running buffer

Dilute 1:20 from commercially available 20× buffer (Invitrogen)

Step 7 Proteinase K storage buffer

Component	Final concentration	Stock	Amount
Proteinase K	20 mg ml^{-1}		200 mg
Tris–HCl, pH 8	50 mM	1 M	0.5 ml
$CaCl_2 \cdot 2H_2O$	30 mM	1 M	30 μl
Glycerol	50%	100%	5 ml
H_2O to 10 ml			

2× Proteinase K buffer

Component	Final concentration	Stock	Amount
Tris–HCl, pH 7.5	100 mM	1 M	1 ml
EDTA–NaOH, pH 8.0	12.5 mM	0.5 M	0.25 ml
NaCl	150 mM	3 M	0.5 ml
SDS	2% (v/v)	20%	1 ml
H_2O			7.25 ml

Acidic Phenol/Chloroform/IAA (25:24:1)

Combine 25-ml acidic phenol, 24-ml chloroform and 1-ml isoamyl alcohol (overlay with 0.1-M citrate buffer, pH 4.3 ± 0.2 which you can take from the acidic phenol bottle)

Step 8 50% DMSO

Mix 1-ml DMSO with 1-ml H_2O

10× RNA ligase buffer without ATP

Component	Final concentration	Stock	Amount
Tris–HCl, pH 7.6	0.5 M	1 M	5 ml
$MgCl_2 \cdot 6H_2O$	0.1 M	1 M	1 ml
2-Mercaptoethanol	0.1 M	14.3 M	0.07 ml
Acetylated BSA	1 mg ml^{-1}	20 mg ml^{-1}	0.5 ml
H_2O			3.43 ml

2× Formamide loading dye

Component	Final concentration	Stock	Amount
EDTA–NaOH, pH 8.0	50 mM	0.5 M	2 ml
Bromophenol blue	0.05% (w/v)		5 mg
Formamide			8 ml

10× TBE

Component	Final concentration	Stock	Amount
Tris base	445 mM		53.9 g
Boric acid	445 mM		27.5 g
EDTA–NaOH, pH 8.0	10 mM	0.5 M	20 ml
H_2O to 1 l			

0.4 M NaCl

Component	Final concentration	Stock	Amount
NaCl	0.4 M	3 M	66.7 ml
H_2O			433.3 ml

Step 9 10× RNA ligase buffer with ATP

Component	Final concentration	Stock	Amount
Tris–HCl, pH 7.6	0.5 M	1 M	5 ml
$MgCl_2 \cdot 6H_2O$	0.1 M	1 M	1 ml
2-Mercaptoethanol	0.1 M	14.3 M	0.07 ml
Acetylated BSA	1 mg ml^{-1}	20 mg ml^{-1}	0.5 ml
ATP	2 mM	100 mM	0.2 ml
H_2O			3.23 ml

Step 10 10× dNTP solution

Component	Final concentration	Stock	Amount
dATP	2 mM	0.1 M	0.2 ml
dCTP	2 mM	0.1 M	0.2 ml
dGTP	2 mM	0.1 M	0.2 ml
dTTP	2 mM	0.1 M	0.2 ml
H_2O			9.2 ml

150-mM KOH/20-mM Tris base

Component	Final concentration	Stock	Amount
KOH	150 mM	5 M	30 μl
Tris base	20 mM	1 M	20 μl
H_2O			950 μl

150-mM HCl

Component	Final concentration	Stock	Amount
HCl, concentrated	150 mM	12.1 M	12.4 μl
H_2O			987.6 μl

Step 11 10× PCR buffer

Component	Final concentration	Stock	Amount
Tris–HCl, pH 8.0	100 mM	1 M	1 ml
KCl	500 mM	1 M	5 ml

2-Mercaptoethanol	10 mM	14.3 M	7 µl
Triton X-100	1% (v/v)	100%	0.1 ml
$MgCl_2 \cdot 6H_2O$	20 mM	1 M	2 ml
H_2O			1.9 ml

5× DNA loading dye

Component	**Final concentration**	**Stock**	**Amount**
EDTA–NaOH, pH 8.0	50 mM	0.5 M	1 ml
Bromophenol blue	0.2% (w/v)		20 mg
Ficoll type 400	20% (w/v)		2 g
H_2O to 10 ml			

Step 12 1 M DTT

Dissolve 1.54-g DTT in 10-ml water

3 M NaOAc (pH 5.2)

Component	**Stock**	**Final concentration**	**Amount**
NaOAc	n/a	n/a	246.09 g
H_2O to 1 l			

HPLC buffer A

Component	**Final concentration**	**Stock**	**Amount**
Acetonitrile	3%	100%	30 ml
TEAA	0.1 M	2 M	50 ml
H_2O to 1 l			

HPLC buffer B

Mix 900-ml acetonitrile with 100-ml water

4. PROTOCOL

4.1. Preparation

4.1.1 Cells

Expand cells in an appropriate growth medium containing selection antibiotics as appropriate to maintain your stable cell line. We usually prepare lysates from 3 to 5 ml of wet cell pellet from crosslinked cells per experiment. This corresponds to 20–50 15-cm cell culture plates (for HEK293). However, if material is limiting, we have performed successful PAR-CLIPs experiments from <0.5 ml of wet cell pellet (200 × 10^6 HEK293 cells (10 15-cm plates) will yield ~1 ml of wet cell pellet).

Grow cells to ~80% confluence. Fourteen hours before crosslinking, add 4-thiouridine (4SU) to a final concentration of 100 μM directly to the cell culture medium. 6-Thioguanosine (6SG, 100 μM) can also be used as the photoactivatable ribonucleoside. Induce expression of protein, if necessary.

4.2. Tip

If you want to add 4SU to 50 15-cm cell culture plates containing 20 ml of growth medium each, place 265 ml of growth medium into a sterile, empty bottle (e.g., an empty media bottle). Add 132.5-μl 1 M 4SU and mix. Additional reagents such as doxycycline (e.g., 1 μg ml^{-1}) to induce protein expression may be added. Dispense 5 ml of the prepared growth medium containing 4SU per 15-cm plate.

4.2.1 Buffers

Buffer recipes and required reagents are listed above. Allow ~1 day for general preparations, including buffer preparation. All pH measurements and adjustments are performed at room temperature. Buffers and all perishable reagents should be refrigerated for storage. We use water purified by a Millipore water purification system.

On the day before you start the PAR-CLIP procedure, fill the required amounts of the individual buffers into 50-ml conical tubes and refrigerate them. The table below gives a rough guide to the required amounts of the individual buffers (but only of those that will be used in quantities above 1 ml on the first 2 days). Add DTT and protease inhibitors on the day of the experiment.

Buffer	Amount per sample
PBS	About 1 l for 20–30 15-cm cell culture plates
Citrate–phosphate buffer	5 ml
NP40 lysis buffer	3 ml per ml cell pellet volume
IP wash buffer	3 ml + 1/10 cell pellet volume
High-salt wash buffer	3 ml
Phosphatase wash buffer	2 ml
PNK buffer without DTT	7 ml

4.2.2 Antibodies

This protocol was originally developed using anti-FLAG antibodies; use of a different antibody will likely require the optimization of the appropriate IP and wash conditions prior to starting a large-scale experiment. Ensure that the optimal salt concentration for antibody binding is maintained throughout the protocol; washing the immunoprecipitate with high salt may disrupt antibody–antigen interactions.

If in doubt, you can use NP40 lysis buffer instead of the IP and the high-salt wash buffers; however, removal of nonspecifically interacting RNAs might be less efficient.

The table below shows the protocol modifications for an anti-AGO2 antibody (Millipore, 04-642) that we have also used successfully:

Step	Anti-FLAG antibody	Anti-AGO2 antibody
Wash after IP	3× IP wash buffer	3× NP40 lysis buffer
KCl concentration	*300 mM*	*150 mM*
High-salt wash	3× high-salt wash buffer	3× NP40 lysis buffer
KCl concentration	*500 mM*	*150 mM*

Except for these two buffers, all the other washing steps are performed as described.

4.2.3 Radiolabeling

On day 3, you will need 5′-^{32}P-radiolabeled RNA size markers. It is advisable to prepare them before starting the PAR-CLIP experiment. Perform a standard radiolabeling procedure with T4 PNK and [γ-^{32}P]-ATP according to the manufacturer's guidelines and gel-purify the markers (e.g., phosphorylate 1-μM RNA size marker in a 10-μl reaction volume using 1 μl of [γ-^{32}P]-ATP (see RNA Radiolabeling). Keep radioactive gel pieces from the running front of this gel as markers to implant into gels for alignment of phosphorimager printouts to exposed gels later on.

4.3. Caution

Consult your institute's Radiation Safety Officer for proper ordering, handling, and disposal of radioactive materials.

4.4. Duration

Preparation	Expanding cell line(s)	Approximately 2 weeks depending on the desired number of cells
	Antibody testing	Variable
	Buffers etc.	1 day
	Radiolabeling of RNA size markers	1.5 days
Protocol	Total	7 days
	Day 1	3–4 h
	Day 2	10–12 h
	Day 3	5–6 h
	Day 4	3–4 h
	Day 5	5–6 h
	Day 6	6–7 h
	Day 7	4–5 h

See Figure 8.1 for the flowchart of the complete protocol.

5. STEP 1 UV CROSSLINKING OF 4-THIOURIDINE-LABELED CELLS (DAY 1)

5.1. Overview

In this first step, the RNA-binding protein of interest is cross-linked to its bound mRNAs targets that incorporated the photoactivatable ribonucleoside into nascent transcripts during the labeling step (see also UV crosslinking

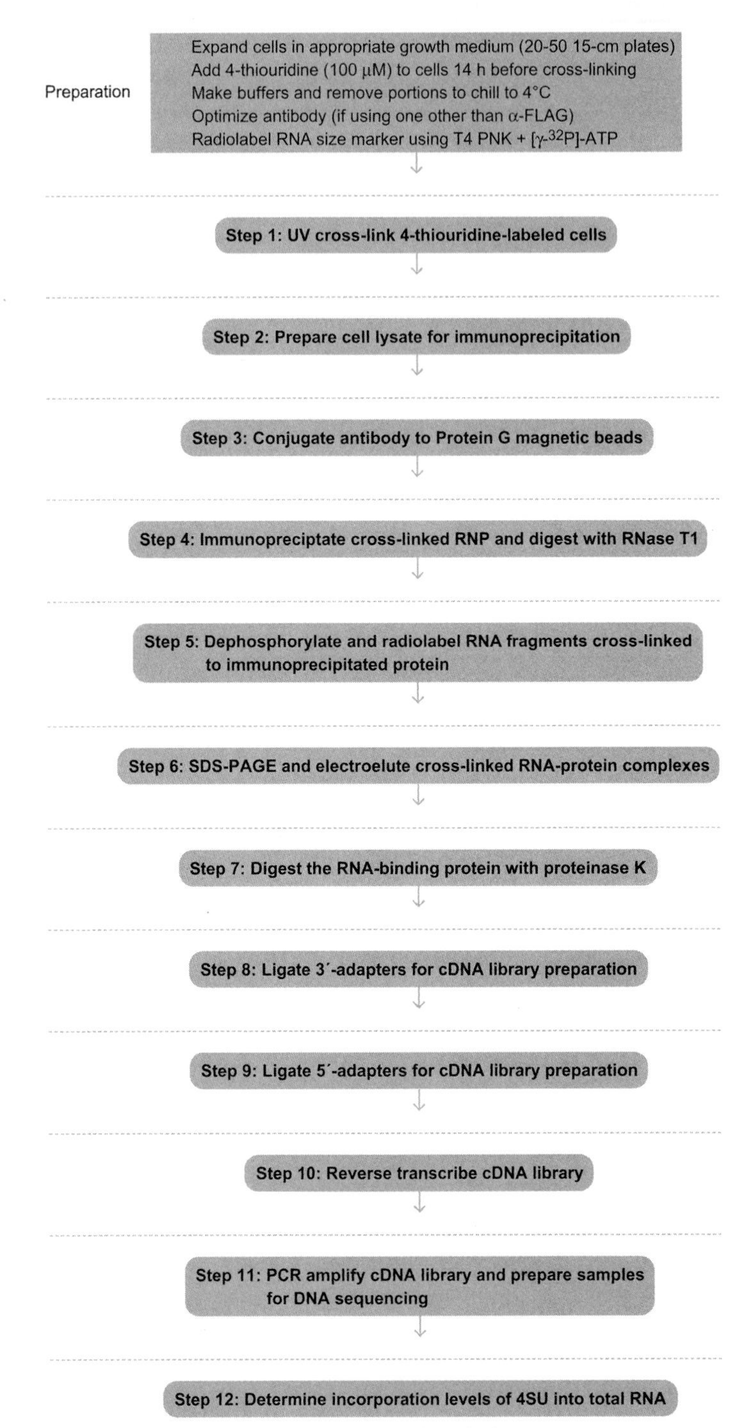

Figure 8.1 Flowchart of the complete protocol, including preparation.

of interacting RNA and protein in cultured cells). The cells are then collected and the resulting cell pellet will be used as the input for the following PAR-CLIP procedure.

5.2. Duration

About 2–3 h

5.2.1 For adherent cells

1.1a Decant growth medium.

1.2a Wash cells once with 5-ml ice-cold PBS per plate and remove PBS completely by decanting and inverting the cell culture dish.

1.3a Place plates on a tray filled with ice to keep cells cold and irradiate uncovered with 0.15 J cm^{-2} total energy of 365-nm UV light in a Stratalinker 2400 or similar device.

1.4a Add 3-ml PBS per plate and dislodge cells with a cell scraper. Transfer to prechilled 50-ml centrifugation tubes on ice. After the cells from the last plate have been collected, centrifuge at 500 × *g* for 5 min at 4 °C and discard the supernatant. Expect to obtain about 5 ml of wet cell pellet from 50 15-cm plates.

1.5a (*Optional: Pause point*) If you do not want to proceed directly to the cell lysis, snap- freeze the cell pellet in liquid nitrogen and store at 80 °C. Cell pellets can be stored for at least 12 months.

5.2.2 For cells grown in suspension culture

1.1b Collect cells by centrifugation at 500 × *g* for 5 min at 4 °C.

1.2b Wash cells by resuspending in 20-ml ice-cold PBS and spin again at 500 × *g* for 5 min at 4 °C.

1.3b Resuspend cells in 20-ml ice-cold PBS and transfer into one 15-cm cell culture plate.

1.4b Place plate on a tray with ice and irradiate uncovered with 0.2 J cm^{-2} of 365-nm UV light in a Stratalinker 2400 or similar device.

1.5b Transfer cells into a 50-ml centrifugation tube, collect by centrifugation at 500 × *g* for 5 min at 4 °C, and discard the supernatant.

1.6b (*Optional: Pause point*) If you do not want to proceed directly to the cell lysis, snap-freeze the cell pellet in liquid nitrogen and store at −80 °C. Cell pellets can be stored for at least 12 months.

5.3. Tip

Keep cell suspensions on ice until centrifugation.

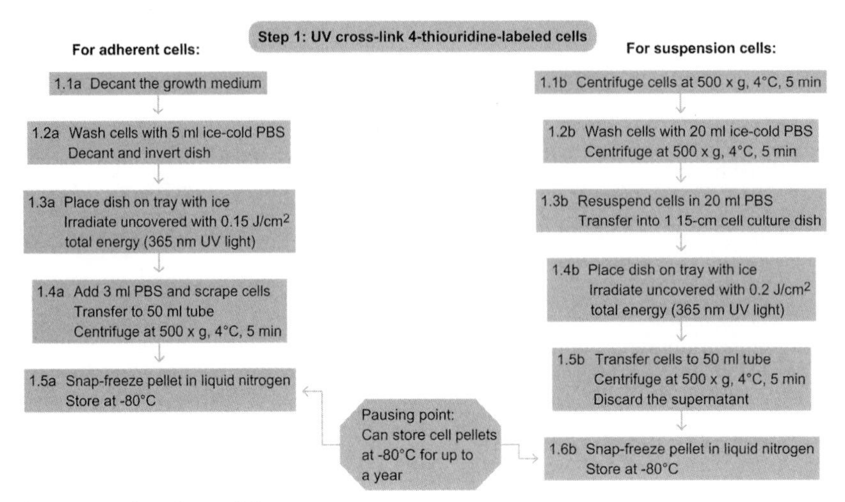

Figure 8.2 Flowchart of Step 1.

See Fig. 8.2 for the flowchart of Step 1.

6. STEP 2 PREPARATION OF CELL LYSATE FOR IMMUNOPRECIPITATION (DAY 2)

6.1. Overview

The cell pellet obtained on day 1 will be lysed (see also Lysis of mammalian and Sf9 cells) in preparation for immunoprecipitation. Partial RNase T1 digestion of mRNAs facilitates the recovery of cross-linked mRNPs.

6.2. Duration

2 h

2.1 Thaw the cross-linked cell pellet on ice. Prepare the magnetic beads (see Step 3) while the pellet thaws. Resuspend the cell pellet in 3 cell pellet volumes of NP40 lysis buffer and incubate on ice for 10 min.

2.2 Clear the cell lysate by centrifugation at 13 000 × *g* for 15 min at 4 °C.

2.3 Clear the lysate further by filtering it through a 5-μm membrane syringe filter. Attach the syringe filter to a 20-ml syringe, remove the plunger, and transfer the supernatant into the syringe. Be careful to hold the syringe above the 50-ml conical tube since the lysate will start to drip through the filter by gravity at first. Insert the plunger and gently apply pressure until all of the lysate is filtered. Depending on the initial viscosity of the lysate, it might be necessary to exchange a clogged filter for a fresh one.

2.4 Add RNase T1 to a final concentration of 1 U μl^{-1} and incubate in a water bath for 15 min at 22 °C. Invert to mix from time to time. Cool reaction for 5 min on ice before proceeding.

2.5 Remove a 10-μl aliquot for immunoblotting as a control for the protein levels used as input and freeze at −20 °C.

6.3. Tip

Take the cell pellet out of the −80 °C freezer and put it on ice first thing in the morning since the thawing process takes a long time.

6.4. Tip

The temperature and duration of the incubation with RNase T1 are both critical for obtaining a controlled partial digestion.

6.5. Tip

Designate a set of micropipettors for working with RNases to avoid contamination at later RNA isolation and cDNA library preparation steps.

See Fig. 8.3 for the flowchart of Step 2.

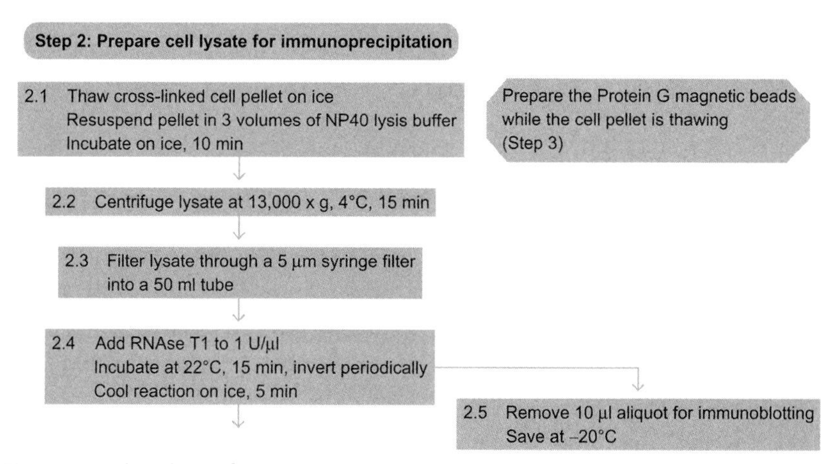

Figure 8.3 Flowchart of Step 2.

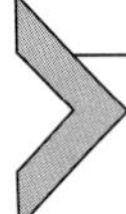

7. STEP 3 PREPARATION OF THE MAGNETIC BEADS (DAY 2)

7.1. Overview

The antibody is conjugated to protein G magnetic beads to be used in the subsequent immunoprecipitation. Protein G is the optimal Ig-binding protein for anti-FLAG antibodies based on species and isotype. The choice of protein A versus protein G should be considered depending on the antibody used.

7.2. Duration

1.5 h

3.1 Transfer 10 μl of Protein G magnetic particles per ml cell lysate (typically ~100–150 μl of beads) to a 1.5-ml microtube. Put the magnetic rack on ice. Wash the beads twice with 1 ml of citrate–phosphate buffer.

3.2 Resuspend the beads in twice the volume of citrate–phosphate buffer relative to the original volume of bead suspension (i.e., 200–300 μl).

3.3 Add antibody to a final concentration of 0.25 mg ml^{-1} and incubate on a rotating wheel for 40 min at room temperature.

3.4 Collect the beads and wash twice in 1 ml of citrate–phosphate buffer to remove unbound antibody.

3.5 Resuspend beads in twice the volume of citrate–phosphate buffer relative to the original volume of bead suspension.

7.3. Tip

This step is performed while the cell pellet is thawing.

7.4. Tip

Be careful to not let the magnetic beads dry out.

See Fig. 8.4 for the flowchart of Step 3.

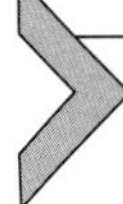

8. STEP 4 IMMUNOPRECIPITATION AND SECOND RNase T1 Treatment (Day 2)

8.1. Overview

The mRNA-RBP complex of choice is isolated from the lysate by immunoprecipitation. A second RNase T1 digestion ensures that only the RNA

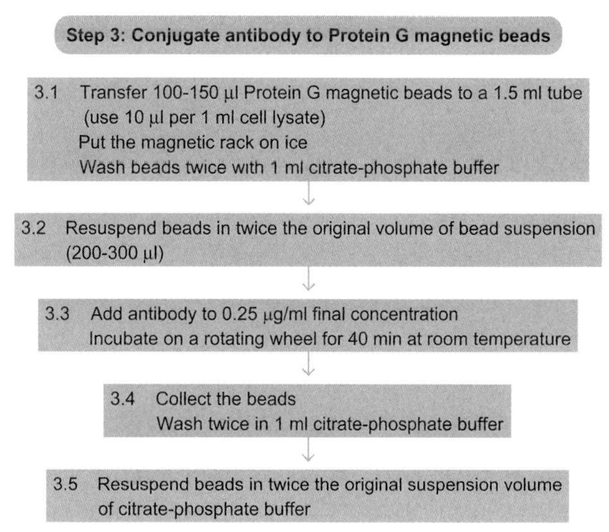

Figure 8.4 Flowchart of Step 3.

segment that was bound, crosslinked, and protected by the RBP is recovered and sequenced. This enables the precise definition of the binding sites.

8.2. Duration

2 h

4.1 Add 20 µl of freshly prepared antibody-conjugated magnetic beads per ml of the partially RNase T1-treated cell lysate from Step 2 and incubate in a 15-ml centrifuge tube on a rotating wheel for 1 h at 4 °C.

4.2 Collect magnetic beads on a magnetic particle collector for 15-ml centrifuge tubes (Invitrogen) and remove the supernatant. Save an aliquot for immunoblotting.

4.3 Add 1 ml of IP wash buffer and transfer to 1.5-ml polypropylene tubes.

4.4 Wash beads 2 times in 1 ml of IP wash buffer.

4.5 Resuspend beads in the original bead volume of IP wash buffer.

4.6 Add RNase T1 (Fermentas, 10 000 U μl^{-1}) to a final concentration of 100 U μl^{-1} and incubate the bead suspension in a 22 °C water bath for 15 min. Cool on ice for 5 min.

4.7 Wash beads 3 times with 1-ml high-salt wash buffer.

4.8 Resuspend beads in 1 volume of dephosphorylation buffer.

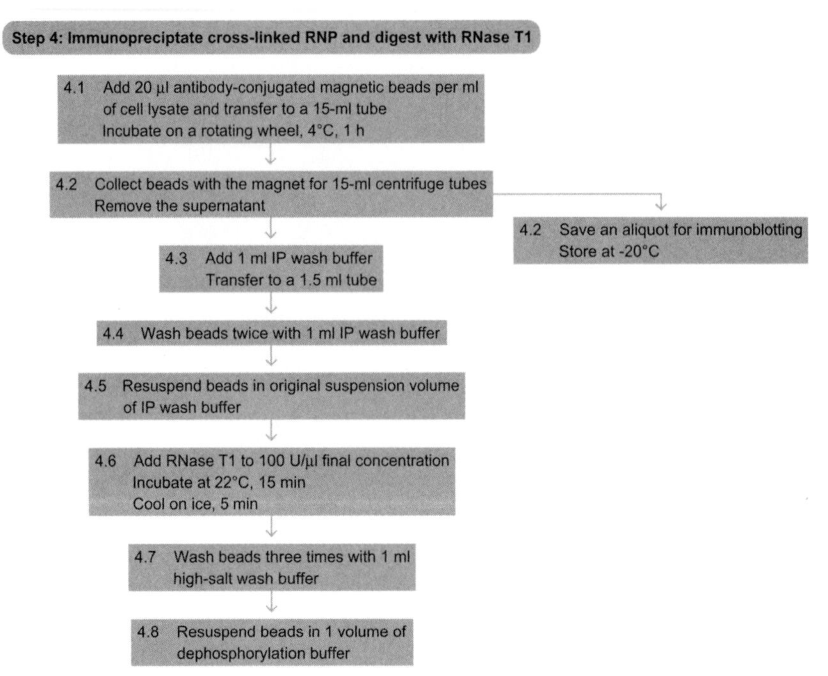

Figure 8.5 Flowchart of Step 4.

8.3. Tip

The RNase T1 incubation temperature and time are both crucial to avoid overdigestion of RNA. Overdigestion could result in RNA segments that are too short to be mapped uniquely to transcript or genomic sequences.

See Fig. 8.5 for the flowchart of Step 4.

9. STEP 5 DEPHOSPHORYLATION AND RADIOLABELING RNA SEGMENTS CROSSLINKED TO IMMUNOPRECIPITATED PROTEINS (DAY 2)

9.1. Overview

The RNAs crosslinked by the RBP of interest are radiolabeled using T4 PNK and [γ-^{32}P]-ATP in order to visualize them by autoradiography after fractionation by SDS-PAGE (see also RNA Radiolabeling).

9.2. Duration

2 h

5.1 Add calf intestinal alkaline phosphatase (CIP from NEB) to a final concentration of 0.5 U μl^{-1}, and incubate the suspension for 10 min at 37 °C and mixing at 800 rpm.

5.2 Wash beads twice with 1 ml of phosphatase wash buffer.

5.3 Wash beads twice with polynucleotide kinase (PNK) buffer without DTT.

5.4 Resuspend beads in one original bead volume of PNK buffer containing DTT.

5.5 Add [γ-^{32}P]-ATP to a final concentration of 0.1 μCi μl^{-1} and T4 PNK (NEB) to 1 U μl^{-1}. Incubate the suspension for 30 min at 37 °C and 800 rpm, mixing manually every 5–10 min.

5.6 Add 100-μM nonradioactive ATP and incubate for another 5 min at 37 °C. This ensures that all RNAs are fully 5′ phosphorylated, which is required for the ligation of the 5′ adapter (Step 9).

5.7 Wash the magnetic beads 5 times with 800 μl of PNK buffer without DTT; dispose of the radioactive buffer according to local guidelines.

5.8 Resuspend the beads in 65 μl of 1× SDS-PAGE loading buffer, incubate for 5 min in a heat block at 90 °C to denature, and release the immunoprecipitated RBP with the cross-linked, radiolabeled RNAs from the beads. Vortex.

5.9 Remove the magnetic beads on the separator and transfer the supernatant to a clean 1.5-ml microcentrifuge tube. (*Pause point*: you can freeze the supernatant and continue with the protocol at another time.)

9.3. Tip

Remove the [γ-^{32}P]-ATP from the freezer and place it at room temperature during the dephosphorylation incubation so that it is thawed by the time you need it.

See Fig. 8.6 for the flowchart of Step 5.

10. Step 6 SDS-PAGE and Electroelution of Cross-Linked RNA-Protein Complexes from Gel Slices (Days 2 and 3)

10.1. Overview

Size fractionation of the radiolabeled and cross-linked RNA protein complexes is achieved by SDS-PAGE (see One-dimensional SDS-Polyacrylamide Gel Electrophoresis (1D SDS-PAGE)). The band corresponding to the expected mass of the protein will be excised and the cross-linked RNA protein complexes electroeluted. This step ensures that only the band corresponding to the correct RBP is isolated and additionally prevents any unbound but labeled RNA from further processing (see Fig. 8.7(a)).

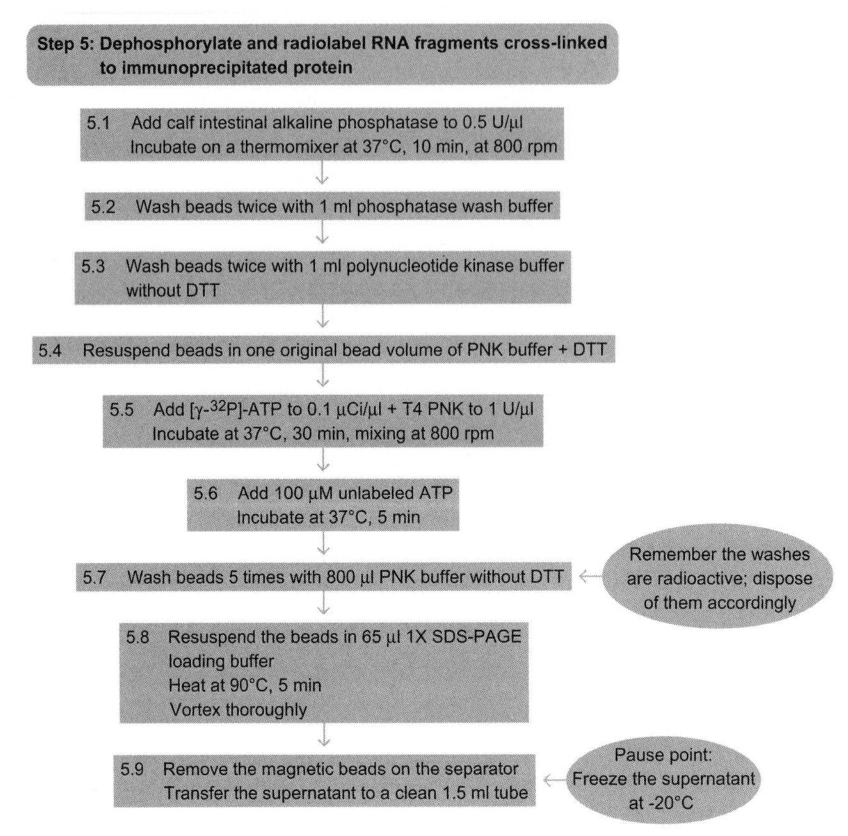

Figure 8.6 Flowchart of Step 5.

10.2. Duration

4.5 h

6.1 Load 2 × 30 μl of the supernatant into two adjacent wells on a Novex Bis-Tris 4–12% (Invitrogen) precast SDS-PAGE gel. Leave at least one empty lane between different samples or different proteins of interest. Load a protein ladder on both sides of the gel. Save the remaining 5 μl of the bead eluate for immunoblotting.

6.2 Run the gel in 1× MOPS-SDS running buffer for 45–60 min at 200 V until the loading dye has reached the bottom of the gel.

6.3 Disassemble the gel chamber (the buffer will be radioactive!) and carefully dismantle the gel, leaving it mounted on one plate. Cut the protruding bottom of the gel so that the gel will lie flat on the phosphorimager screen.

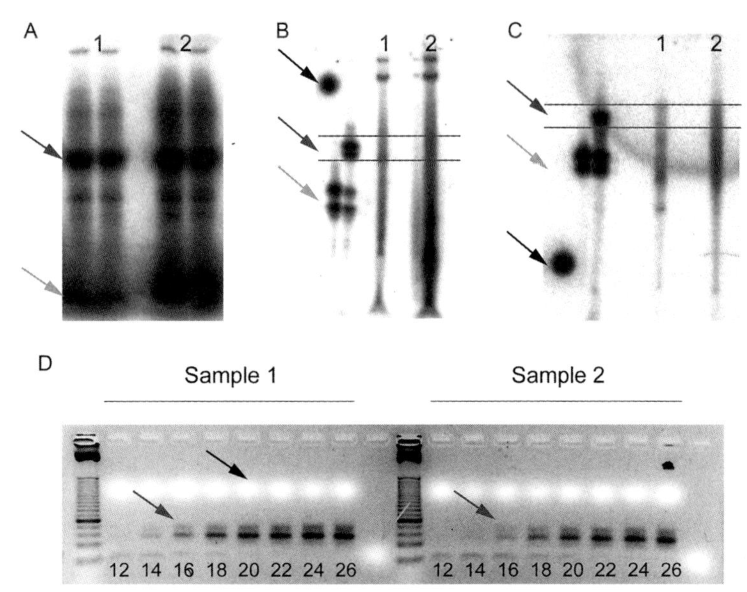

Figure 8.7 Selected PAR-CLIP experimental steps. (a) SDS-PAGE gel of cross-linked and 5′-radiolabeled RNA-protein complex immunoprecipitates. The red arrow points to the radioactive bands corresponding to the expected size of the RNA-binding protein (FUS, running at 75 kDa), and the yellow to the radioactive running front. (b) 8 M urea polyacrylamide gel after 3′-adapter ligation. The black arrow indicates one of the inserted little radioactive gel pieces to facilitate alignment of the gel to printout, the red to the 3′-ligated size markers, and the area that was cut from the gel and further processed; the yellow arrows show the unligated 3′-size markers. (c) 8-M urea polyacrylamide gel after 5′-adapter ligation. The black arrow indicates one of the inserted little radioactive gel pieces to facilitate alignment of the gel to printout, the red to the 5′-ligated size markers, and the area which was cut from the gel and further processed; the yellow arrows show the unligated 5′-size markers. (d) Agarose gel after small-scale trial PCR. The black arrow points to the position of migration of the xylene cyanol loading dye, and the yellow to the bromophenol blue loading dye running close to the gel front. Bands of about 75 and 100 bp are detectable, representing insert-less 5′-adapter-3′-adapter PCR side product and expected insert-containing PCR product, respectively. The red arrows indicate the number of cycles chosen for the large-scale experiment. A 25-bp ladder is loaded to the left of each set of experiments; the fourth band from the bottom corresponds to 100 bp. The negative control was performed but is not shown. (For interpretation of the references to color in this figure legend, the reader is referred to the online version of this chapter.)

6.4 To facilitate the alignment of the gel to the phosphorimager paper printout later on, place three tiny radioactive gel pieces asymmetrically into three of the four corners of the gel. Radioactive gel pieces could be collected earlier from the bottom of the gel from radiolabeling the size markers (see above).

6.5 After placing the gel pieces, wrap the gel in plastic wrap and expose the gel to a phosphorimager screen for 15 min. Visualize it on a phosphorimager. Have a second screen ready and expose it during the scanning process should the first exposure indicate that a longer exposure is necessary.

6.6 Print the scanned image file at its original size (100%). Align the transparently wrapped gel on top of the printout guided by the implanted gel pieces for precise positioning. Cut out the bands that correspond to the expected size of the RBP (see Fig. 8.7(a)).

6.7 Add 800 μl of water to a D-Tube Dialyzer Midi Tube used for electroelution and let stand at room temperature for 5 min. Remove the water. Take care not to pierce the membrane.

6.8 Transfer the excised bands to the dialyzer tube and add 800-μl 1× MOPS-SDS running buffer.

6.9 Place the electroelution rack with the tubes in an agarose gel chamber so that the membrane is exposed to the flow of the current (for details, see manufacturer's instructions). Add 1× MOPS-SDS running buffer to the chamber until it covers the tubes.

6.10 Electroelute the cross-linked RNA-RBP complex at 100 V for 1.5 h. Reverse the current for 2 min to release any protein attached on the dialysis membrane.

6.11 Transfer the solution to two siliconized tubes so that each contains around 350 μl (you will not be able to fully recover the original 800 μl). (*Pause point*: you can freeze the solution at −20 °C and continue the next day.)

10.3. Tip

To confirm that the correct band was excised from the gel, run a second small-scale SDS-PAGE gel with 1 or 2 μl of the remaining 5 μl of your sample (see above). After transferring the gel to a nitrocellulose membrane, take an autoradiography exposure (exposing for 1–2 h) and then use protein-specific antibodies to perform a standard immunoblot (see Western Blotting using Chemiluminescent Substrates). After overlaying the resulting images, you should be able to establish which band corresponds to your protein of interest and proceed with the protocol.

10.4. Tip

In case you observe more than one band, you can also cut all of them since they should correspond to other co-purifying RNA cross-linked proteins.

10.5. Tip

Make sure that the membrane of the dialyzer tube is aligned correctly to allow flow of current.

10.6. Tip

Use aerosol barrier tips and take general precautions to avoid any RNase contamination since you will be working with RNA from now on until the reverse transcription step on day 6. Clean your pipettors before starting to work with RNA.

10.7. Tip

Use siliconized tubes until you have obtained your cDNA library; at low concentrations, nucleic acids have a tendency to stick to the tube walls.

See Fig. 8.8 for the flowchart of Step 6.

11. STEP 7 PROTEINASE K DIGESTION (DAY 3)

11.1. Overview

In this step, the recovered RBP is proteolyzed and the cross-linked RNA is recovered so that it can serve as the input material for subsequent ligation to adapters and Solexa sequencing.

11.2. Duration

3.5 h

7.1 Add 1 volume of 2× Proteinase K Buffer, followed by the addition of Proteinase K (Roche) to a final concentration of 1.2 mg ml^{-1}. Incubate at 55 °C for 30 min. If the volume per tube exceeds 800 μl at this stage, split the sample into two tubes.

7.2 Add 1 volume of acidic phenol/chloroform/isoamyl alcohol, vortex and spin at 20 000 × *g* for 10 min at 4 °C. Recover the upper aqueous phase without disturbing the interphase and pipette into two siliconized tubes.

7.3 Add an equal volume of chloroform, and vortex and spin at 20 000 × *g* for 10 min at 4 °C. Recover the aqueous phase without disturbing the interphase.

7.4 Add 1/10 volume 3 M NaCl, 1 μl of glycogen (10 mg ml^{-1} stock), and 3 volumes of 100% ethanol.

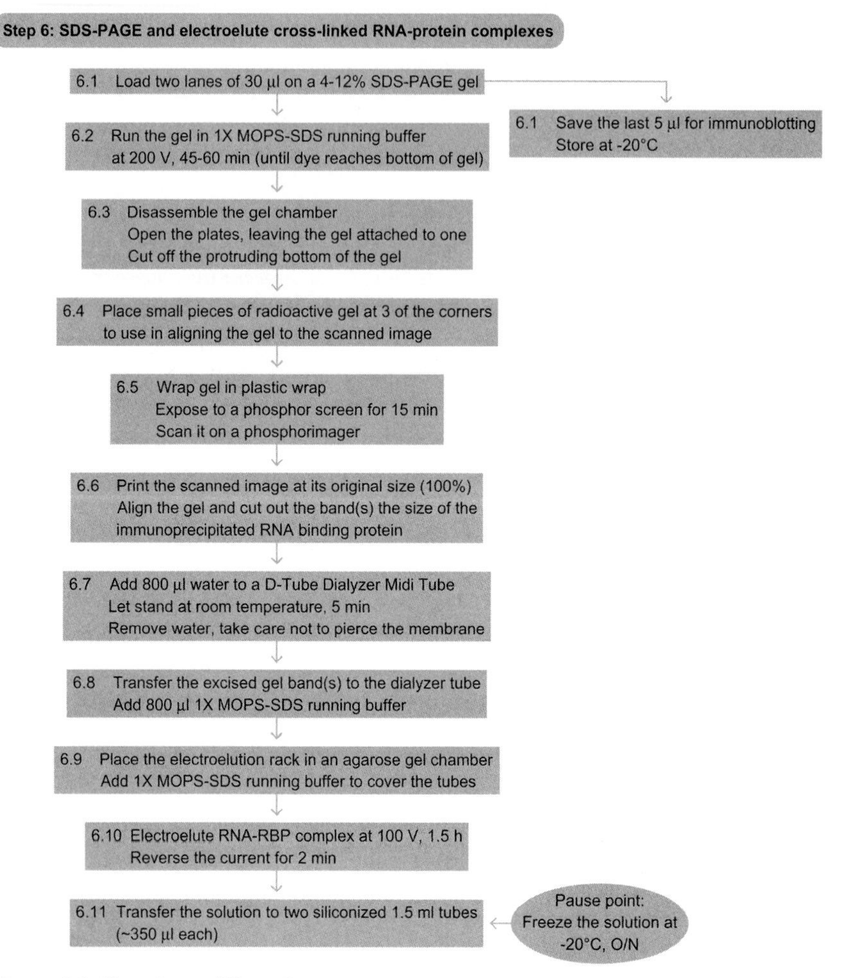

Figure 8.8 Flowchart of Step 6.

7.5 Precipitate the RNA for 1 h on ice and spin at 20 000 × *g* for 15 min at 4 °C. (*Pause point*: precipitate the RNA overnight at 20 °C. See also RNA purification – precipitation methods.)

7.6 Remove the supernatant, air-dry the pellets, and dissolve in a total of 10 µl of H_2O.

11.3. Tip

Monitor the radioactivity of the supernatant and the pellet to assess the efficiency of the ethanol precipitation.

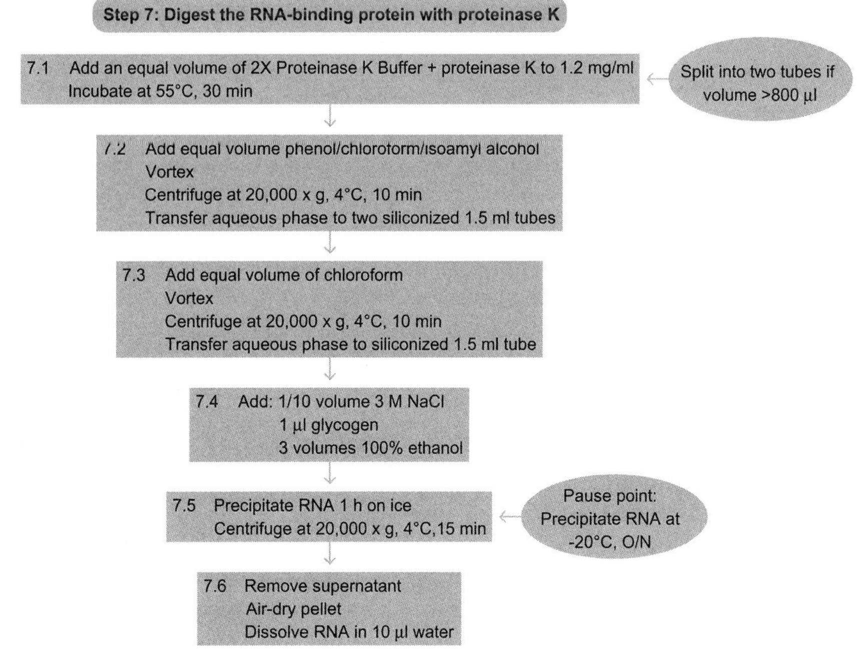

Figure 8.9 Flowchart of Step 7.

11.4. Tip

Repeat the phenol/chloroform/IAA extraction until there is no precipitate visible in the interphase (usually, once is sufficient, but two or more times might be needed).

See Fig. 8.9 for the flowchart of Step 7.

12. STEP 8 3′-ADAPTER LIGATION FOR cDNA Library Preparation (Day 3 overnight, day 4, beginning of day 5)

12.1. Overview

The recovered 5′-^{32}P-phosphorylated RNA is now carried through a standard cDNA library preparation protocol, originally described for the cloning of small regulatory RNA (Hafner et al., 2008). As a first step, a preadenylated 3′-adapter is ligated using T4 Rnl2(1-249)K227Q ligase (see Fig. 8.7(b)).

12.2. Duration

Day 3: 30 min (+ overnight incubation)

Day 4: 4–5 h (highly dependent on required exposure time)

Day 5: 2 h

8.1 Prepare the following reaction mixture for ligating the 3′-adapter, multiplying the volumes by the number of ligation reactions plus two (one for the size markers plus another to account for pipetting loss):

2 μl of 10× RNA ligase buffer (without ATP)
6 μl 50% DMSO
1 μl of 100 μM preadenylated 3′-adapter
Add 9 μl of the reaction mixture to each sample (from Step 7.6, you have 19 μl per tube).

8.2 Prepare ~40 fmol of a 1:100 dilution of 5′-^{32}P-labeled RNA size markers (19-nt and 24-nt size marker at equimolar concentrations, see preparation step). This controls for successful ligation and indicates the length of the bands that will be cut out from the gel later on.

8.3 Heat-denature the RNA to disrupt secondary structures by incubating at 90 °C for 30 s. Immediately place the tubes on ice for 30 s.

8.4 Add 1 μl of Rnl2(1-249)K227Q ligase (1 μg μl^{-1}) to the ligation reactions, mix gently, and incubate overnight on ice in the cold room or in an insulated ice bucket covered with a lid.

8.5 The next morning, cast a 15% denaturing 8-M urea polyacrylamide gel (we use the UreaGel system from National Diagnostics. See also RNA purification by preparative polyacrylamide gel electrophoresis) and wait until the polymerization process is complete. Our gels measure 15 × 17 cm × 0.8 mm and contain about 25-ml gel volume with a 20 well comb.

8.6 Prerun the gel in 1× TBE buffer at 30 W for 30 min. After the prerun, flush the wells with 1× TBE.

8.7 Add 20 μl of formamide gel loading solution to the samples to stop the ligation reactions.

8.8 Denature the RNA at 90 °C for 30 s.

8.9 Load each sample into one well (or two) of the gel. Load the size markers symmetrically on both sides of the gel to allow for approximating the length of the ligated samples between them. Use the center lanes of the gel to guarantee even running of the gel. Make sure to space different samples appropriately, typically at a two-well distance, to avoid cross contamination. Ensure that the overall loading of the gel is asymmetrical.

8.10 Run the gel at 30 W for 45 min until the bromophenol blue dye is close to the bottom of the gel.

8.11 Dismantle the gel, leaving it attached to one glass plate. To facilitate the alignment of the gel to the phosphorimager paper printout, again implant three tiny radioactive gel pieces asymmetrically at three of the four corners of the gel. Cover the gel with plastic wrap.

8.12 Expose the gel to a phosphorimager screen for at least 1 h. If the radioactivity of the recovered RNA is weak, you can expose the gel overnight, placing the exposure cassette in a −20 °C freezer. Allow the cassette to return to room temperature before opening it.

8.13 Align the gel on top of a full-scale printout according to the position of the three radioactive gel pieces. Cut out the bands in between the ligated products of the 19-nt and above the 24-nt marker (*Note*: We do not recommend cutting of RNA that is running below the 19-nt marker line. For our bioinformatic analyses, all sequences shorter than 20 nucleotides are discarded due to the increased probability of mapping to multiple locations and the uncertainty defining its genetic location. Our bioinformatic analysis pipeline discards reads under 20 nucleotides for that reason. In case you would like to cut a larger size range, we also have successfully used two larger-sized markers (33-nt and 35-nt). Cut out the ligated 19- and 24-nt size markers; they will serve once more as a ligation control in the next step (see Figure 8.7(b)).

8.14 Place the cut gel pieces in siliconized tubes and add 350-µl 0.4-M NaCl (ensure that the gel pieces are covered by NaCl). Elute the ligation products overnight at 4 °C, shaking at 800 rpm.

8.15 Transfer the supernatant into a new siliconized tube and add 1-ml 100% ethanol. Precipitate the RNA for 1 h on ice and spin at 20 000 × *g* for 15 min at 4 °C.

8.16 Remove the supernatant, air-dry the pellets, and dissolve in a total of 9-µl H_2O. Dissolve the ligated markers in 12-µl H_2O.

12.3. Tip

Keep the supernatant from the ethanol precipitation. In case no pellet should form after the precipitation, you can add 1 µl of glycogen to it and precipitate again. We do not routinely add glycogen at this stage, since the relatively high amount of glycogen might interfere with the subsequent reaction, which is performed in a small volume. The linear acrylamide eluted from the gel is usually a sufficient carrier.

See Fig. 8.10 for the flowchart of Step 8.

Figure 8.10 Flowchart of Step 8.

13. STEP 9 5′-ADAPTER LIGATION FOR cDNA Library Preparation (Day 5, beginning of day 6)

13.1. Overview

In this step, the 5′-adapter is joined to the 3′-ligated RNA to enable the cDNA synthesis in the next step (see Fig. 8.7(c)).

13.2. Duration

Day 5: about 5 h (highly dependent on required exposure time)
Day 6: 2 h

9.1 Prepare the following reaction mixture for the ligation of the 5′-adapter, multiplying the volumes by the number of ligation reactions to be performed plus two (for the positive control plus one extra to account for pipetting loss):

1 μl of 100 μM 5′-adapter
2 μl of 10× RNA ligase buffer with ATP
6 μl 50% DMSO

Combine 9 μl of this mixture with 9 μl of sample.

Remember to also process the ligated markers from the last step. Ligate 9 μl out of the 12 μl and keep 3 μl as an unligated control for the next gel.

9.2 Denature the RNA by incubation at 90 °C for 30 s. Immediately place the tube on ice for 30 s.

9.3 Add 2 μl of T4 RNA ligase 1 (10 U μl^{-1}), mix gently, and incubate for 1 h at 37 °C.

9.4 In the meantime, cast a 12% denaturing 8-M urea polyacrylamide gel and wait until the polymerization process is complete. We again use 0.8-mm spacers and a 20 well comb.

9.5 Prerun the gel 1× TBE buffer at 30 W for 30 min. After the prerun, gently flush the wells with 1× TBE.

9.6 Add 20 μl of formamide gel loading solution to the samples, incubate them at 90 °C for 30 s, and load the gel. Make sure to space different samples appropriately, typically at a two-well distance, to avoid cross contamination.

Load 50% of the ligated markers on the left side and 50% on the right side. Load the remaining 3-μl unligated marker on one side (see Fig. 8.7(c)).

9.7 Run the gel at 30 W for 45 min until the bromophenol blue dye is close to the bottom of the gel. Disassemble and image the gel as described above for the 3′-ligation (start with an exposure roughly twice as long as for the 3′-ligation) and excise the new ligation product (also excise the ligated markers).

9.8 Elute the ligation products from the gel slices in 350-μl 0.4-M NaCl. Shake at 800 rpm overnight at 4 °C. Add 1 μl of 100 μM 3′ PCR primer as a carrier to facilitate the recovery of the ligation products.

9.9 Pipet the supernatant into a new siliconized tube and add 1-ml 100% ethanol. Precipitate the RNA for 1 h on ice and spin at 20 000 × *g* at 4 °C for 15 min.

9.10 Remove the supernatant, air-dry the pellets, and dissolve in 5.6-μl H_2O.

13.3. Tip

Make sure that the loading of the gel is asymmetrical.

13.4. Tip

You can recover unligated material by excising the gel region below the ligated 19-nt marker line, since this represents 3′-ligated, 5′-unligated RNA fragments. Freeze these gel pieces. You will have them stored as a backup in case you later wish to perform another 5′-adapter ligation from the RNAs eluted from these gel pieces.

See Fig. 8.11 for the flowchart of Step 9.

14. STEP 10 cDNA Library Preparation/Reverse Transcription (Day 6)

14.1. Overview

The RNA ligated to both sequencing adapters is reverse-transcribed and will be used for PCR in the subsequent step.

14.2. Duration

1.5 h

10.1 Prepare the following reaction mix (multiplied by the number of samples plus one for pipetting loss):
 1.5-μl 0.1-M DTT
 3-μl 5× First-strand buffer (Superscript)
 4.2-μl 10× dNTPs

10.2 Denature the RNA by incubating the tube at 90 °C for 30 s and transfer the tube to a 50 °C thermomixer.

10.3 Add 8.7 μl of the reaction mix to each sample and incubate at 50 °C for 3 min. Add 0.75 μl of Superscript III Reverse Transcriptase and incubate at 42 °C for 1 h.

10.4 Prepare 150-mM KOH/20-mM Tris base and 150-mM HCl and use pH paper to verify that a 1:1 mixture results in a pH between 7.0 and 9.5. If not, change the ratios until the pH is within that range.

10.5 To hydrolyze the RNA, add 40 μl of 150-mM KOH/20-mM Tris base and incubate at 90 °C for 10 min.

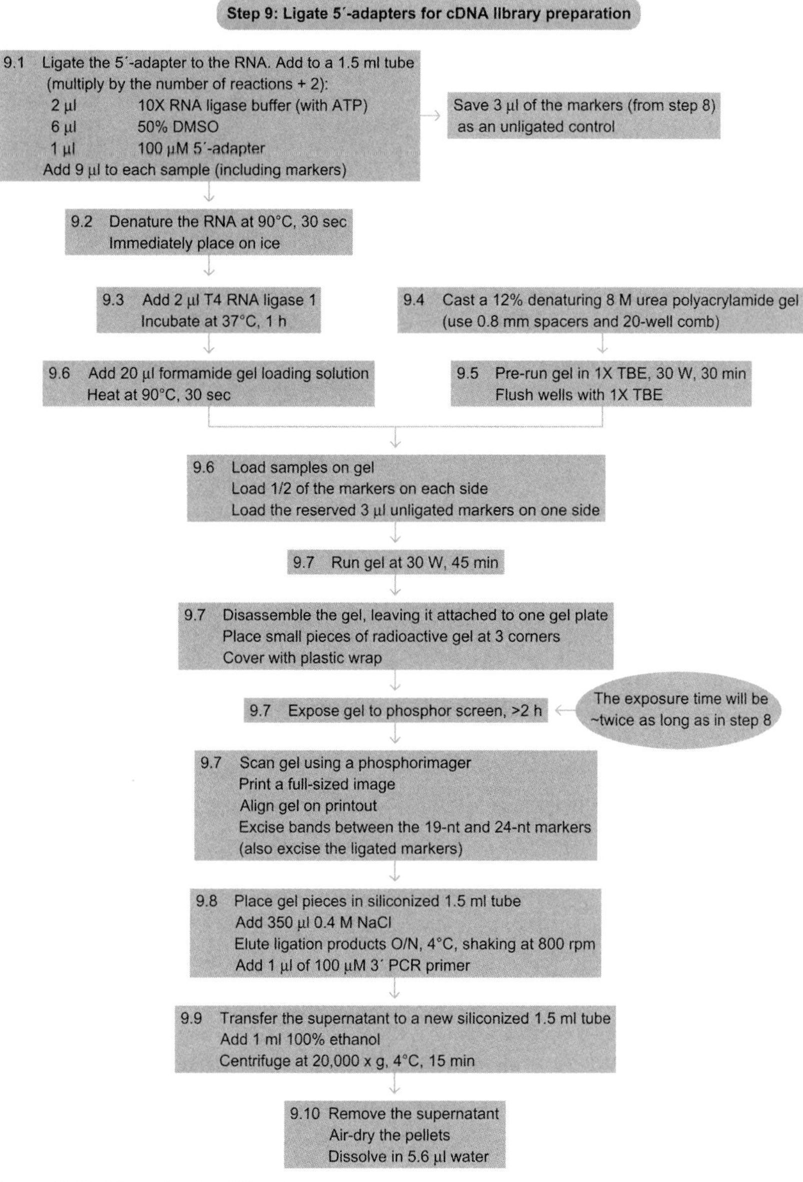

Figure 8.11 Flowchart of Step 9.

10.6 Neutralize the solution by adding 40 μl of 150-mM HCl (the exact volume depends on the ratio determined in Step 10.4) and check the pH of the mixture by spotting 1 μl on pH paper. It should be between 7.0 and 9.5 so that the subsequent PCR is not inhibited. If necessary, readjust the pH by adding more base or acid.

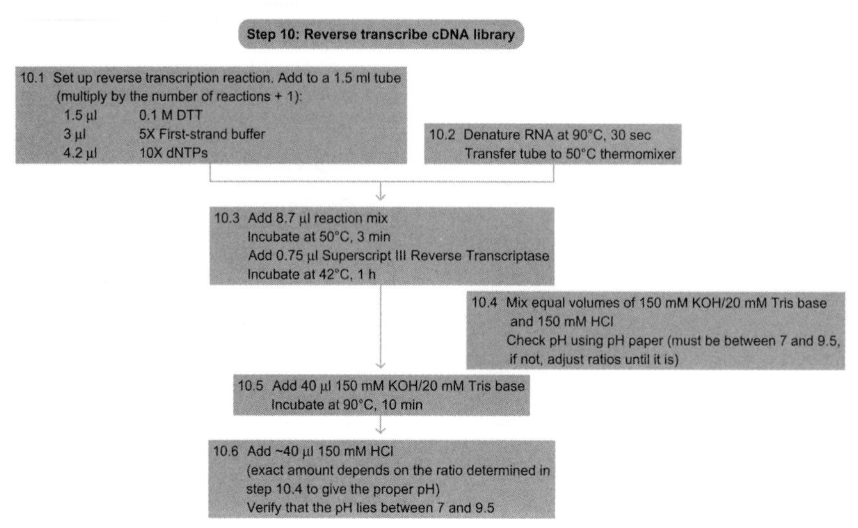

Figure 8.12 Flowchart of Step 10.

See Fig. 8.12 for the flowchart of Step 10.

15. STEP 11 PCR AMPLIFICATION OF cDNA Library & Sample Preparation for Sequencing (Day 6)

15.1. Overview

This step concludes the PAR-CLIP protocol. To minimize the distortion of the cDNA library composition by excessive PCR and to recognize possible failure during reverse transcription leading to false positive PCR results, we monitor the accumulation of the PCR product during a pilot PCR. To determine the minimal cycle number, a small-scale trial PCR is performed before the final large-scale PCR. The PCR product is gel fractionated (see Agarose Gel Electrophoresis); the appropriately sized fraction is recovered from the gel and submitted to Solexa sequencing (see Fig. 8.7(d) and Explanatory Chapter: Next Generation Sequencing).

15.2. Duration

8–9 h

11.1 Prepare the following mix multiplied by the number of samples (plus one for the negative control):

40-μl 10× PCR buffer

40-μl 10× dNTPs

2 μl of 100-μM 5′ PCR primer
2 μl of 100-μM 3′ PCR primer
272-μl H_2O

89 μl of the reaction mix will be used in the pilot PCR reaction to determine the minimal cycle number; the remainder will be needed for the large-scale PCR (freeze the reaction mix if you want to run the large-scale PCR on the following day).

Add to a 0.2-ml thin-walled PCR tube:

89 μl of the reaction mix
10-μl cDNA
1-μl *Taq* polymerase (5 U μl^{-1})

Include a negative control using water instead of cDNA.

Use the following cycling conditions:

94 °C 45 s
50 °C 85 s
72 °C 60 s

11.2 To determine the necessary number of cycles for amplifying the cDNA library, remove 12-μl aliquots every other cycle starting with cycle number 12 and ending with cycle number 26. To remove aliquots from the PCR tube, temporarily pause the PCR cycler at the end of the 72 °C step. You can use a multichannel pipettor to remove the aliquots.

11.3 Analyze 6 μl of each sample on a 2.5% agarose gel containing 0.4 $\mu g\ ml^{-1}$ of ethidium bromide to check for consistency. Load a 25-bp ladder on each side and load all cycles from one sample next to one another in an ascending order.

The PCR products might appear as a double band, with the higher band running at the expected length of about 95–110 nucleotides and a lower band corresponding to the 3′-adapter-to-5′-adapter ligation/template switch products running at about 65 nucleotides. Figure 8.7(d) illustrates a typical small-scale PCR. The red arrows indicate the chosen number of cycles for the large-scale experiment.

Define the minimal cycle number for the cDNA amplification. It should be within the exponential amplification phase of the PCR, about five cycles away from reaching the saturation level of PCR amplification. For a typical PAR-CLIP experiment, the minimal number of cycles is between 16 and 20 (*Pause point*: you can pause at any time before or after the large-scale PCR).

11.4 For the large-scale PCR, set up three 100-μl reactions (as in Step 11.1). Perform the PCR using the determined minimal number of cycles. After the reactions are finished, combine all three PCR reactions. Include a negative control as before.

11.5 Analyze 6 μl of the products next to the corresponding products from the pilot PCR on a 2.5% agarose gel containing 0.4 μg ml^{-1} of ethidium bromide to check for consistency.

11.6 To the remaining 264 μl, add 26.4-μl 3 M NaCl and 1-ml 100% ethanol. Precipitate for 1 h on ice and spin at 20 000 × *g* at 4 °C for 30 min. Remove the supernatant, air-dry the pellet, and dissolve in 40 μl of 1× DNA loading dye (dilute 5× DNA loading dye in 1× TBE).

11.7 Divide the sample into two wells of a 2.5% low melt agarose gel containing 0.4-μg ml^{-1} ethidium bromide. Run the gel at 120 V for 2–3 h.

Do not overload the gel since that will compromise its ability to separate fragments.

11.8 Visualize the DNA on a 365-nm transilluminator and use a clean scalpel to excise the region corresponding to 95–110 nucleotides.

11.9 Purify the DNA using the Qiaquick gel extraction kit (Qiagen) according to the manufacturer's instructions. Include the isopropranol step as described for short fragments. Elute in 30-μl elution buffer.

11.10 Analyze 5 μl of the eluate on a 2.5% agarose containing 0.4 μg ml^{-1} of ethidium bromide gel to ensure the removal of any unwanted amplified 5′-adapter-3′-adapter PCR products.

11.11 Submit 10 μl of the purified cDNA to Solexa sequencing.

15.3. Tip

If you have more than one PAR-CLIP sample, prepare a 96-well plate with 3-μl 5× DNA loading dye in the required number of wells so that you only have to pipet once per cycle using a multichannel pipettor.

15.4. Tip

The main goal of the preparative gel is to reduce noninformative sequence reads of unwanted 5′-adapter-3′-adapter PCR products. Do not overload the gel and run the gel as long as possible to achieve the best separation possible. Check intermittently so that you do not run your samples into the buffer.

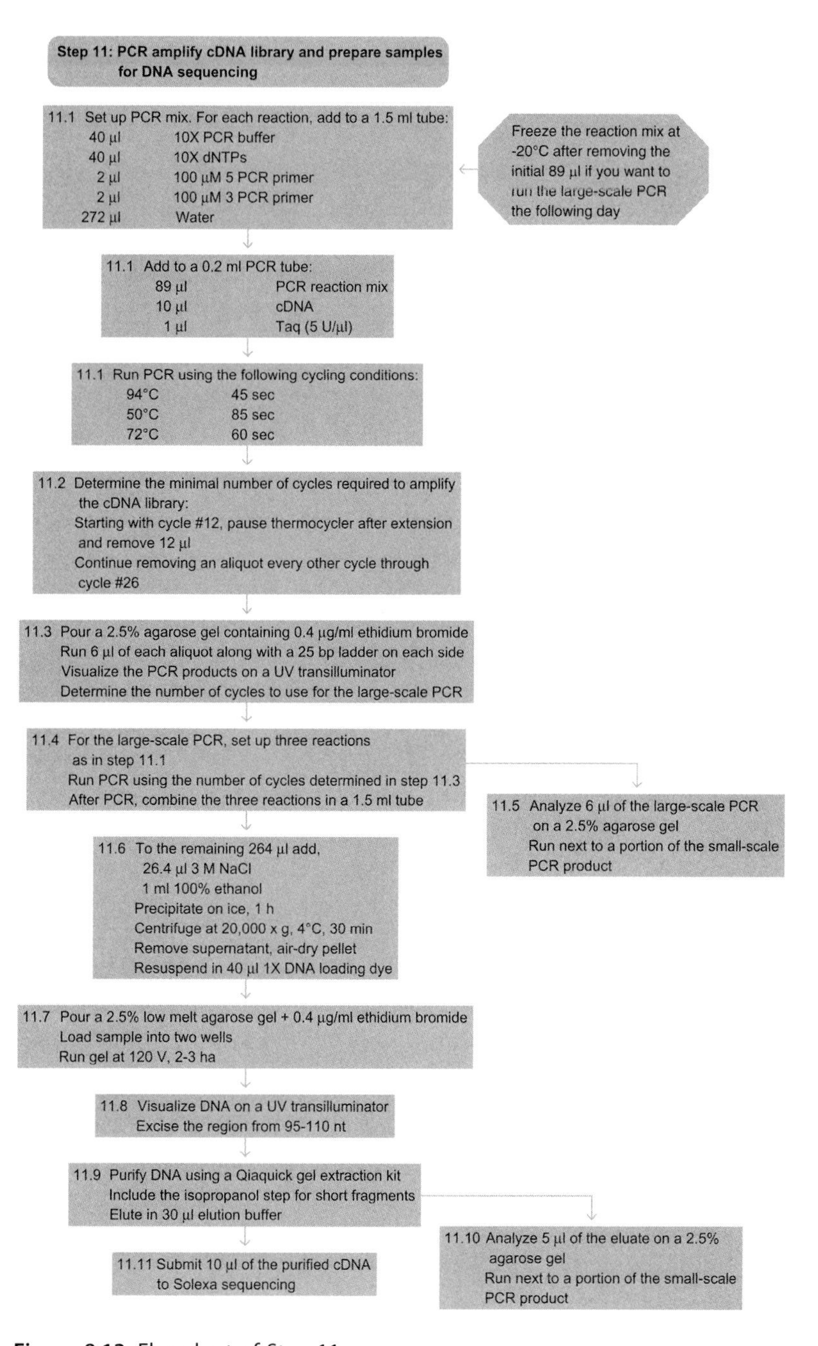

Figure 8.13 Flowchart of Step 11.

15.5. Tip

Bromophenol blue runs roughly at the same position as the samples. Use a DNA gel loading buffer containing xylene cyanol or do not add any dye to it. Run an aliquot containing bromophenol blue next to your samples.

15.6. Tip

Perform a second gel extraction if any 5′-adapter-3′-adapter products are still seen after the first gel extraction.

See Fig. 8.13 for the flowchart of Step 11.

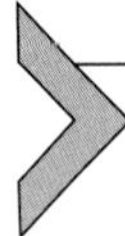

16. STEP 12 DETERMINATION OF INCORPORATION LEVELS OF 4SU INTO TOTAL RNA

16.1. Overview

To optimize crosslinking of protein to RNA, it is useful to determine the fraction of substitution of uridine by 4SU. This is especially necessary when changing cell growth conditions or cell type. Total RNA is isolated and enzymatically degraded to monomeric ribonucleosides, which are separated and quantified by HPLC analysis (Andrus and Kuimelis, 2001).

16.2. Duration

Cell labeling and harvesting: 16 h + 20 min
RNA isolation: about 2 h
Dephosphorylation and hydrolysis: 30 min + overnight incubation
Chromatography: as needed

12.1 Grow two 10-cm plates of HEK293 cells in regular medium. Add 100-μM 4SU to one plate 16 h prior to harvesting cells.
12.2 Decant the growth medium; wash cells once with 1× PBS.
12.3 Add 1 ml of TRIzol reagent directly onto the plate and isolate the total RNA according to the manufacturer's instructions.
12.4 Include 0.1-mM DTT in the isopropanol and ethanol-wash steps as well as in the subsequent reaction to prevent oxidation of the thiocarbonyl group, which yields disulfides or uridine.
12.5 Dissolve the RNA pellet in 60-μl H_2O containing 1-mM DTT.
12.6 Determine the concentration of the obtained RNA. Expect to obtain about 50–100 μg total RNA per 10-cm plate.
12.7 0.2 OD_{260} (8.0 μg RNA) of total RNA is digested and dephosphorylated to single nucleosides for HPLC analysis. Set up the following reaction:

Reagent or solution	Final concentration	Volume
RNA	0.2 OD_{260}	x µl
10-mM DTT	0.1 mM	1.3 µl
1 M $MgCl_2$	13.8 mM	1.8 µl
0.5-M Tris–HCl (pH 7.5)	34.6 mM	9.0 µl
Bacterial Alkaline Phosphatase	1.6 U	x µl
Snake Venom Phosphodiesterase	0.2 U	x µl
H_2O		to 130 µl

12.8 Digest for 16 h at 37 °C. As an additional control, also digest and analyze synthetic RNAs with and without 4SU.

12.9 Add 2.3-µl 100-mM DTT, 4 µl of 3 M NaOAc (pH 5.2), and 100 µl of ice-cold 100% ethanol, incubate on dry ice for 10 min, and centrifuge the sample at 12 500 × *g* for 5 min at 25 °C.

12.10 Collect the supernatant, add 3 µl of 100-mM DTT and 300 µl of ice-cold 100% ethanol, incubate on dry ice for 10 min, centrifuge the sample at 12 500 × *g* for 5 min at 25 °C, and collect the supernatant.

12.11 Evaporate the supernatant in a Speed-vac to complete dryness. If the sample is not completely dried, the retention times during HPLC analysis are affected.

12.12 Dissolve the sample in 50 µl of H_2O, which is the volume of one HPLC injection.

12.13 Separate ribonucleosides on a Supelco Discovery C18 reverse phase column (bonded phase silica 5 µM particles, 250 × 4.6 mm).

12.14 Use an isocratic gradient of 0% B for 15 min, 0–10% B for 20 min, and 10–100% B for 30 min with a 5-min 100% B wash between runs (see Fig. 8.14).

12.15 Calculate the absorption ratios from the known sequence of the reference oligonucleotides. This is used to estimate the incorporation rate for 4SU (in our experiments, between 1.4 and 2.4% of U is substituted by 4SU).

12.16 Confirm U and 4SU retention times by co-injection with standards.

12.17 Calculate the substitution ratio of 4SU by dividing the area under the curve by the extinction coefficients of rU versus 4SU at 260 nm versus 330 nm.

$$\%\text{U substituted by 4SU} = (\text{Area}_{\text{4SU, 330 nm}} / \varepsilon_{\text{4SU, 330 nm}}) \times (\varepsilon_{\text{U, 260 nm}} / \text{Area}_{\text{U, 260 nm}}) \times 100$$

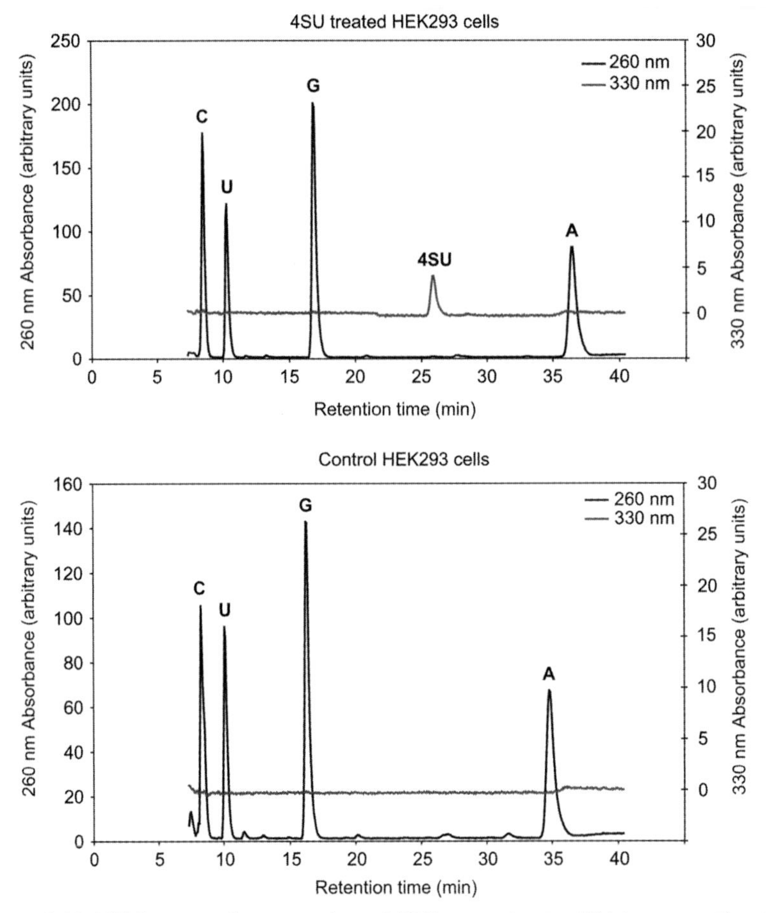

Figure 8.14 HPLC trace of extracted total RNA to estimate 4SU incorporation into HEK293 cells. Please refer to the main text for a detailed description. (For color version of this figure, the reader is referred to the online version of this chapter.)

Nucleoside	Extinction coefficient at 260 nm (pH 7.0)	Extinction coefficient at 330 nm
rA	12 340	0
rC	7020	0
rG	10 240	0
rU	9720	0
4SU	4250	17 000

See Fig. 8.15 for a flowchart of Step 12.

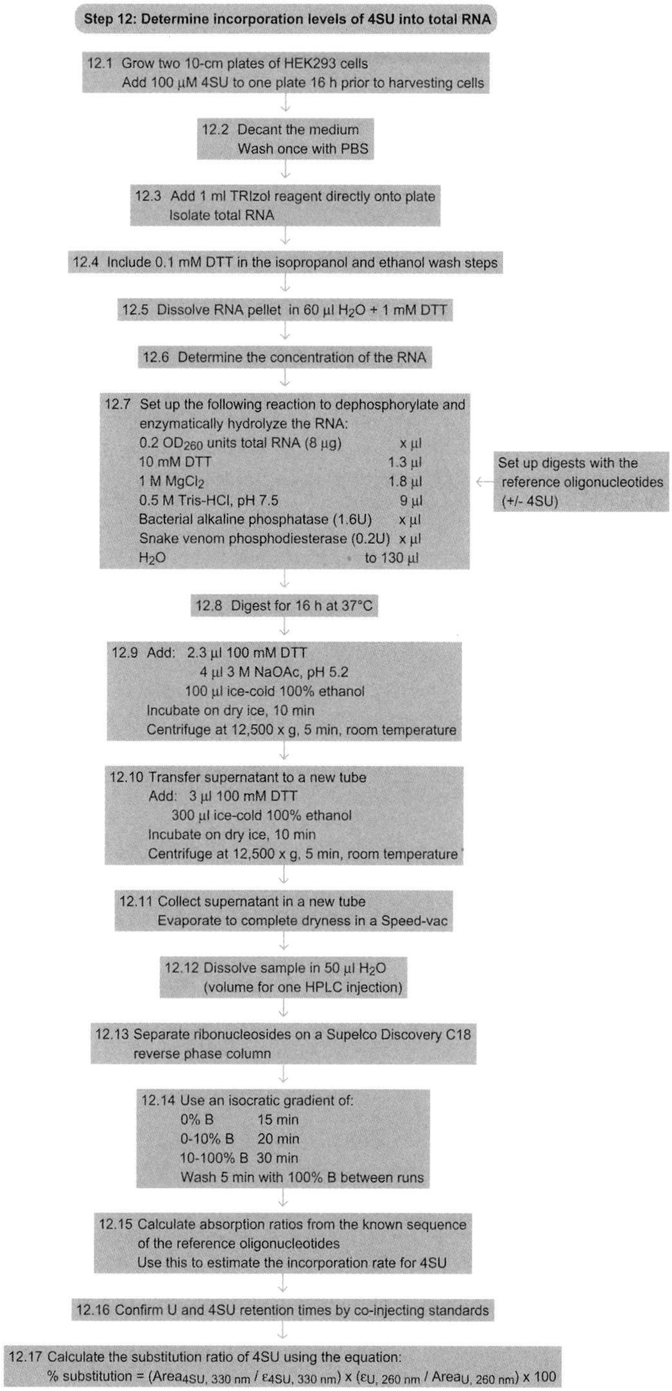

Step 12: Determine incorporation levels of 4SU into total RNA

12.1 Grow two 10-cm plates of HEK293 cells
Add 100 μM 4SU to one plate 16 h prior to harvesting cells

↓

12.2 Decant the medium
Wash once with PBS

↓

12.3 Add 1 ml TRIzol reagent directly onto plate
Isolate total RNA

↓

12.4 Include 0.1 mM DTT in the isopropanol and ethanol wash steps

↓

12.5 Dissolve RNA pellet in 60 μl H_2O + 1 mM DTT

↓

12.6 Determine the concentration of the RNA

↓

12.7 Set up the following reaction to dephosphorylate and enzymatically hydrolyze the RNA:

0.2 OD_{260} units total RNA (8 μg)	x μl
10 mM DTT	1.3 μl
1 M $MgCl_2$	1.8 μl
0.5 M Tris-HCl, pH 7.5	9 μl
Bacterial alkaline phosphatase (1.6U)	x μl
Snake venom phosphodiesterase (0.2U)	x μl
H_2O	to 130 μl

← Set up digests with the reference oligonucleotides (+/- 4SU)

↓

12.8 Digest for 16 h at 37°C

↓

12.9 Add: 2.3 μl 100 mM DTT
4 μl 3 M NaOAc, pH 5.2
100 μl ice-cold 100% ethanol
Incubate on dry ice, 10 min
Centrifuge at 12,500 x g, 5 min, room temperature

↓

12.10 Transfer supernatant to a new tube
Add: 3 μl 100 mM DTT
300 μl ice-cold 100% ethanol
Incubate on dry ice, 10 min
Centrifuge at 12,500 x g, 5 min, room temperature

↓

12.11 Collect supernatant in a new tube
Evaporate to complete dryness in a Speed-vac

↓

12.12 Dissolve sample in 50 μl H_2O
(volume for one HPLC injection)

↓

12.13 Separate ribonucleosides on a Supelco Discovery C18 reverse phase column

↓

12.14 Use an isocratic gradient of:

0% B	15 min
0-10% B	20 min
10-100% B	30 min

Wash 5 min with 100% B between runs

↓

12.15 Calculate absorption ratios from the known sequence of the reference oligonucleotides
Use this to estimate the incorporation rate for 4SU

↓

12.16 Confirm U and 4SU retention times by co-injecting standards

↓

12.17 Calculate the substitution ratio of 4SU using the equation:
$\%\ substitution = (Area_{4SU,\ 330\ nm} / \varepsilon_{4SU,\ 330\ nm}) \times (\varepsilon_{U,\ 260\ nm} / Area_{U,\ 260\ nm}) \times 100$

Figure 8.15 Flowchart of Step 12.

VIDEO

Please refer to this link (http://www.jove.com/index/Details.stp?ID=2034) for a video illustrating the first day of experiments.

REFERENCES

Referenced Literature

Ambros, V. (2004). The functions of animal microRNAs. *Nature*, *431*(7006), 350–355.

Andrus, A., & Kuimelis, R. G. (2001). Base composition analysis of nucleosides using HPLC. In A. Andrus & R. G. Kuimelis (Eds.), *Current Protoccols Nucleic Acid Chemistry*. New York: Wiley Unit 10 16.

Bezerra, R., & Favre, A. (1990). In vivo incorporation of the intrinsic photolabel 4-thiouridine into *Escherichia coli* RNAs. *Biochemical and Biophysical Research Communications*, *166*(1), 29–37.

Favre, A., Moreno, G., Blondel, M. O., Kliber, J., Vinzens, F., & Salet, C. (1986). 4-Thiouridine photosensitized RNA-protein crosslinking in mammalian cells. *Biochemical and Biophysical Research Communications*, *141*(2), 847–854.

Gerber, A. P., Luschnig, S., Krasnow, M. A., Brown, P. O., & Herschlag, D. (2006). Genome-wide identification of mRNAs associated with the translational regulator PUMILIO in *Drosophila melanogaster*. *Proceedings of the National Academy of Sciences of the United States of America*, *103*(12), 4487–4492.

Glisovic, T., Bachorik, J. L., Yong, J., & Dreyfuss, G. (2008). RNA-binding proteins and post-transcriptional gene regulation. *FEBS Letters*, *582*(14), 1977–1986.

Halbeisen, R. E., Galgano, A., Scherrer, T., & Gerber, A. P. (2008). Post-transcriptional gene regulation: From genome-wide studies to principles. *Cellular and Molecular Life Sciences*, *65*(5), 798–813.

Hafner, M., Landgraf, P., Ludwig, J., et al. (2008). Identification of microRNAs and other small regulatory RNAs using cDNA library sequencing. *Methods*, *44*(1), 3–12.

Hafner, M., Landthaler, M., Burger, L., et al. (2010). PAR-CLIP – Transcriptome-wide identification of RNA targets and binding sites of RNA-binding proteins. *Cell*, *141*(1), 129–141.

Keene, J. D. (2007). RNA regulons: Coordination of post-transcriptional events. *Nature Reviews Genetics*, *8*(7), 533–543.

Landthaler, M., Gaidatzis, D., Rothballer, A., et al. (2008). Molecular characterization of human Argonaute-containing ribonucleoprotein complexes and their bound target mRNAs. *RNA*, *14*(12), 2580–2596.

Moore, M. J. (2005). From birth to death: The complex lives of eukaryotic mRNAs. *Science*, *309*(5740), 1514–1518.

Tenenbaum, S. A., Carson, C. C., Lager, P. J., & Keene, J. D. (2000). Identifying mRNA subsets in messenger ribonucleoprotein complexes by using cDNA arrays. *Proceedings of the National Academy of Sciences of the United States of America*, *97*(26), 14085–14090.

Ule, J., Jensen, K. B., Ruggiu, M., Mele, A., Ule, A., & Darnell, R. B. (2003). CLIP identifies Nova-regulated RNA networks in the brain. *Science*, *302*(5648), 1212–1215.

SOURCE REFERENCES

Hafner, M., Landgraf, P., Ludwig, J., et al. (2008). Identification of microRNAs and other small regulatory RNAs using cDNA library sequencing. *Methods*, *44*(1), 3–12.

Hafner, M., Landthaler, M., Burger, L., et al. (2010). PAR-CLIP – Transcriptome-wide identification of RNA targets and binding sites of RNA-binding proteins. *Cell*, *141*(1), 129–141.

Referenced Protocols in Methods Navigator

Explanatory Chapter: Next Generation Sequencing.
UV crosslinking of interacting RNA and protein in cultured cells.
RNA Radiolabeling.
Lysis of mammalian and Sf9 cells.
One-dimensional SDS-Polyacrylamide Gel Electrophoresis (1D SDS-PAGE).
Western Blotting using Chemiluminescent Substrates.
RNA purification – precipitation methods.
RNA purification by preparative polyacrylamide gel electrophoresis.
Agarose Gel Electrophoresis.

CHAPTER NINE

Determining the RNA Specificity and Targets of RNA-Binding Proteins using a Three-Hybrid System

Yvonne Y. Koh*, Marvin Wickens†,1

*Microbiology Doctoral Training Program, University of Wisconsin – Madison, Madison, WI, USA
†Department of Biochemistry, University of Wisconsin – Madison, Madison, WI, USA
[1]Corresponding author: e-mail address: wickens@biochem.wisc.edu

Contents

Methods in Enzymology, Volume 539
ISSN 0076-6879
http://dx.doi.org/10.1016/B978-0-12-420120-0.00009-8

Abstract

The three-hybrid system can be used to identify RNA sequences that bind a specific protein by screening a hybrid RNA library with a protein–activation domain fusion as 'bait.' These screens complement biochemical techniques, for example, SELEX, co-immunoprecipitation, and cross-linking experiments (see UV crosslinking of interacting RNA and protein in cultured cells and PAR-CLIP (Photoactivatable Ribonucleoside-Enhanced Crosslinking and Immunoprecipitation): a step-by-step protocol to the transcriptome-wide identification of binding sites of RNA-binding proteins).

1. THEORY

The general strategy of the three-hybrid system is shown in Fig. 9.1. DNA-binding sites are placed upstream of a reporter gene, which has been integrated into the yeast genome. The first hybrid protein consists of a DNA-binding domain linked to an RNA-binding domain. The RNA-binding domain interacts with its RNA-binding site in a bifunctional ('hybrid') RNA molecule. The other part of the RNA molecule interacts

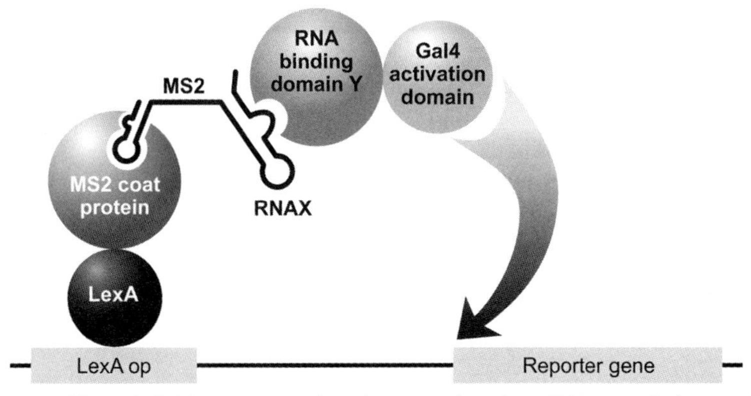

Figure 9.1 Three-hybrid system used to detect and analyze RNA–protein interactions. The diagram depicts the general strategy of the three-hybrid system. Specific protein and RNA components that are typically used are represented. For the sake of simplicity, the following points are not illustrated. In the YBZ-1 strain, both *lacZ* and *HIS3* reporter genes are present under the control of lexA operators (eight in the *lacZ* promoter and four in the *HIS3* promoter). The LexA protein binds as a dimer. The hybrid RNA contains two MS2 stem-loops and the MS2 coat protein binds one stem-loop as a dimer.

with a second hybrid protein consisting of another RNA-binding domain linked to a transcription activation domain (AD). When this tripartite complex forms at the promoter, the reporter gene is turned on. Reporter expression can be detected by phenotype or simple biochemical assays. The specific molecules used most commonly for three-hybrid analysis are depicted in Fig. 9.1. The DNA-binding site consists of a 17-nucleotide recognition site for the *Escherichia coli* LexA protein and is present in multiple copies upstream of both the *HIS3* and *lacZ* genes. The first hybrid protein consists of LexA fused to bacteriophage MS2 coat protein, a small polypeptide that binds to a short stem-loop sequence present in the bacteriophage RNA genome. The hybrid RNA (depicted in more detail in Fig. 9.2) consists of two MS2 coat protein-binding sites linked to the RNA sequence of interest, X. The second hybrid protein consists of the transcription activation domain of the yeast Gal4 transcription factor linked to an RNA-binding protein, Y.

By introducing a hybrid RNA library, cognate partners of a specific RNA-binding protein can be identified. Secondary screens of the initial 'positives' are needed to winnow down candidates for further analysis and to eliminate false positives. Here, we consider a case in which the AD-fusion protein carries a *LEU2* marker and the hybrid RNA vector

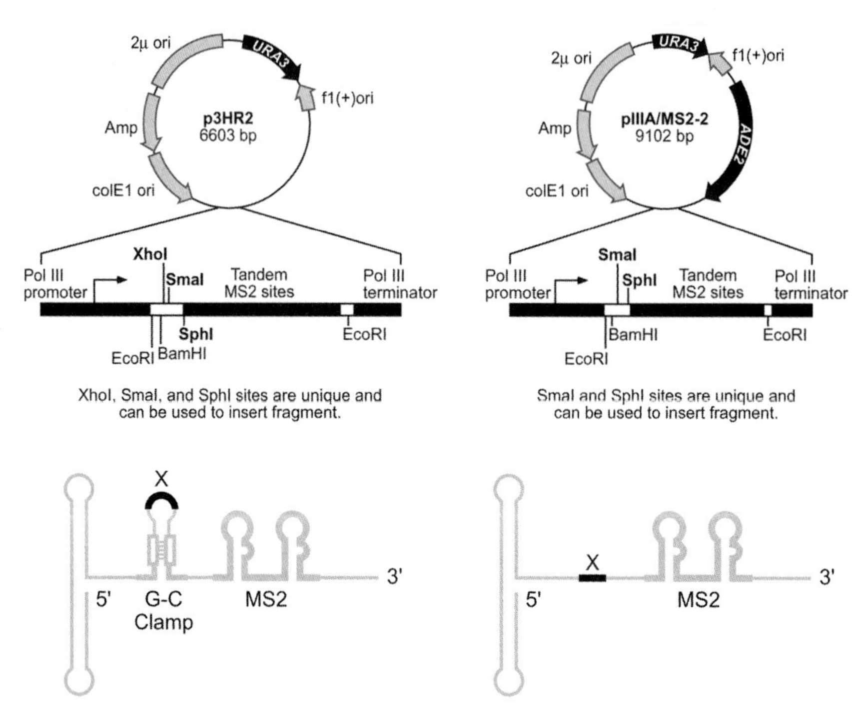

Figure 9.2 Plasmids for the expression of hybrid RNAs. p3HR2 displays the RNA sequence of interest in the loop of a G-C clamp (Stumpf et al., 2008), whereas pIIIA/MS2-2 does not (Zhang et al., 1997). The sequence of interest can be inserted into the unique restriction sites. pIIIA/MS2-2 carries an *ADE2* marker which can be used in Step 4 to eliminate false positives by screening colony color.

(p3HR2) carries the *URA3* marker. When an RNA molecule from the library interacts with the fusion protein, the *HIS3* reporter gene is expressed and the yeast grow on media lacking histidine and/or containing 3-aminotriazole (a competitive inhibitor of the *HIS3* gene product). Additional screens are performed using the *lacZ* reporter and colorimetric assays.

The library selected to analyze in the screen will depend on the purpose of the screen. To identify RNAs that interact with a protein of interest, one can clone expressed sequence tags from a tissue of interest or fragments of a genome (e.g., Sengupta et al., 1999). Alternatively, libraries containing randomized portions of RNA sequences, or portions of a known RNA target, can be used (e.g., Koh et al., 2009).

The output of the system – *HIS3* and *lacZ* activation – has been examined as a function of *in vitro* affinity using a specific RNA–protein interaction (Hook et al., 2005). When the protein FBF-1 was tested with various RNAs, K_d was linearly related to the log of β-galactosidase activity over the range measured, from 1 to 100 nM.

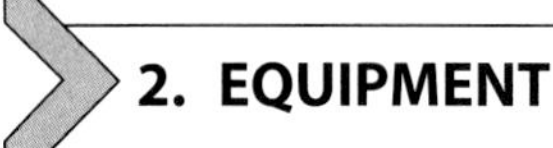

2. EQUIPMENT

Centrifuge
30 °C incubator-shaker
Petri dishes
Glass tubes (18 × 150 mm) with caps
Micropipettors
Micropipettor tips
Nitrocellulose filter (Petri dish-sized)
Whatman 3MM chromatography paper (Petri dish-sized)
Sterile toothpicks
1.5-ml microcentrifuge tubes
Parafilm

3. MATERIALS

Saccharomyces cerevisiae strain YBZ-1: *MATa, ura3-52, leu2-3, -112, his3-200, trp1-1, ade2, LYS2::(LexAop)-HIS3, URA3::(lexAop)-lacZ, and LexA-MS2 MS2 coat (N55K)*
Hybrid RNA library (e.g., p3HR2 or pIIIA/MS2-2)
Activation domain–protein fusion plasmid (e.g., pACTII or pGADT7)
Synthetic dextrose (SD) dropout media (liquid and plates)
3-Aminotriazole (3-AT)
Sodium phosphate dibasic (Na_2HPO_4)
Sodium phosphate monobasic (NaH_2PO_4)
Potassium chloride (KCl)
Magnesium sulfate ($MgSO_4$)
2-Mercaptoethanol
5-Bromo-4-chloro-3-indolyl-β-D-galactopyranoside (X-Gal)
Dimethylformamide
Zymolyase (G-Biosciences)
Wizard Plus SV Miniprep kit (Promega)
Liquid nitrogen

3.1. Solutions & buffers

Step 2 3-Aminotriazole (1 M)

Dissolve 4.21-g 3-AT in 50-ml water. Pass through a 0.2-μm filter to sterilize. Aliquot into tubes, wrap in foil, and store at −50 °C.

Step 4 Z-buffer

Component	Final concentration	Stock	Amount
Na_2HPO_4	60 mM	1 M	60 ml
NaH_2PO_4	40 mM	1 M	40 ml
KCl	10 mM	1 M	10 ml
$MgSO_4$	1 mM	1 M	1 ml
2-Mercaptoethanol	50 mM	14.2 M	Add fresh

Add to 850-ml water. Adjust pH to 7.0, if necessary. Add water to 1 l and filter-sterilize.

X-Gal solution (20 mg ml^{-1})

Dissolve 0.4-g X-Gal in 20-ml dimethylformamide (in a fume hood). Aliquot into tubes, wrap in foil, and store at −20 °C.

4. PROTOCOL

4.1. Preparation

Obtain appropriate hybrid RNA library either from individual laboratories or commercial sources. For preparation of a genomic library, see, for example, SenGupta et al. (1999); for preparation of a randomized library, see, for example, Stumpf et al. (2008).

Transform the AD–protein fusion plasmid into YBZ-1 yeast using the LiAc method (Gietz and Woods, 2002, see also Chemical Transformation of Yeast).

Prepare sufficient plates listed in subsequent steps for the selection (see *Saccharomyces cerevisiae* Growth Media).

4.2. Duration

Preparation	1 week
Protocol	1–2 months

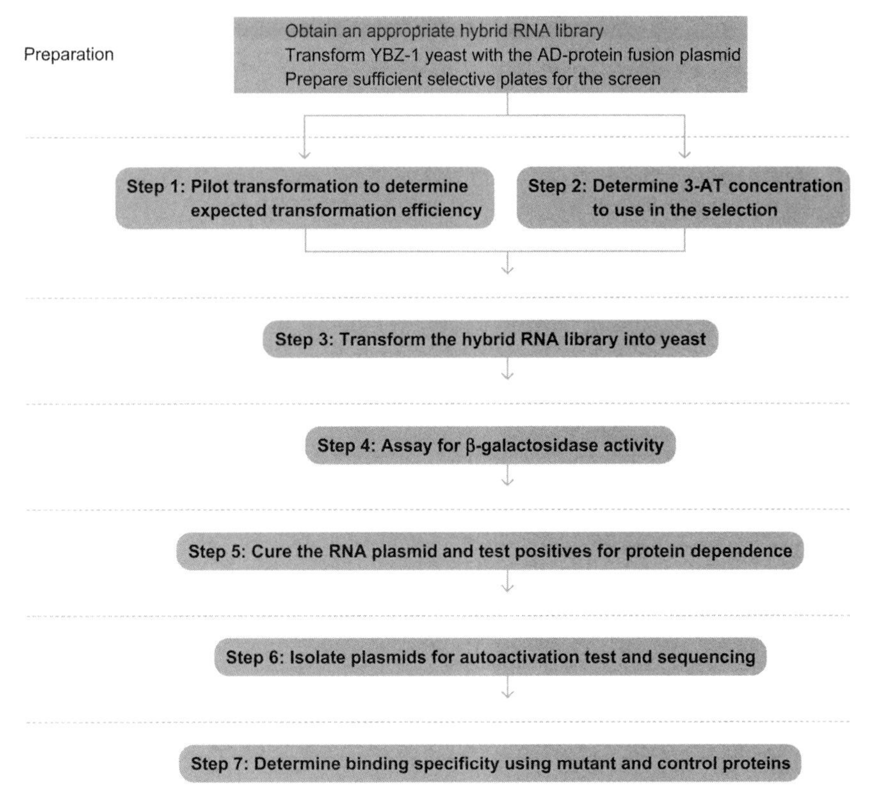

Figure 9.3 Flowchart of the complete protocol, including preparation.

See Fig. 9.3 for the flowchart of the complete protocol.

5. STEP 1 PILOT TRANSFORMATION TO DETERMINE EXPECTED TRANSFORMATION EFFICIENCY

5.1. Overview

Perform a pilot transformation to determine the expected transformation efficiency. This allows one to calculate how much to scale up the transformation in order to obtain the desired number of transformants.

5.2. Duration

3–4 days

1.1 Transform cells from a single YBZ-1 yeast colony carrying the AD–protein fusion plasmid with the hybrid RNA library using the LiAc method (Gietz and Woods, 2002, see Chemical Transformation of

Yeast). Select transformants by plating on SD–Leu–Ura medium. Incubate the plates at 30 °C for 3–4 days.

1.2 Count the number of colonies from a single transformation (~10^8 cells) to determine transformation efficiency (number of transformants/μg DNA/10^8 cells).

1.3 Calculate the number of transformants desired to obtain sufficient coverage of the library. For example, to be 95% confident that a library with 10^6 clones is entirely screened, one must search through 3×10^6 transformants as determined by the following equation:

$$N = \ln(1-p)/\ln(1-(1/u)),$$

where N, transformants; p, probability (% confidence); u, library size.

1.4 Based on the transformation efficiency from Step 1.2, determine the scale of transformation required to obtain the desired number of transformants in Step 1.3.

5.3. Tip

Steps 1 and 2 (pilot transformation and determination of 3-AT concentration to use) can be carried out simultaneously.

5.4. Tip

Yeast can be transformed with both plasmids simultaneously or sequentially. Usually, transformation of the plasmids sequentially yields more transformants. However, transformation of a single plasmid is occasionally toxic to the yeast, requiring a cotransformation of both plasmids.

5.5. Tip

For starters, transform with 0.1–1.0 μg library plasmid DNA. The amount of DNA has to be optimized to obtain a high transformation efficiency.

5.6. Tip

For a detailed yeast transformation protocol, refer to the Gietz Lab webpage (http://home.cc.umanitoba.ca/~gietz/index.html).

See Fig. 9.4 for the flowchart of Step 1.

Step 1: Pilot transformation to determine expected transformation efficiency

1.1 Transform cells from a single YBZ-1 yeast colony carrying the AD-fusion plasmid with the hybrid RNA library using the LiAc method
Plate cells on SD-Leu-Ura plates
Incubate at 30°C, 3-4 days

↓

1.2 Count the number of colonies from a single transformation (~1×10^8 cells)
Determine transformation efficiency: # transformants/µg DNA/10^8 cells

↓

1.3 Calculate the number of transformants needed to give the desired coverage of the library (e.g., to be 95% confident of completely screening a library of 10^6 clones, you must search through 3×10^6 transformants)

↓

1.4 Based on the transformation efficiency, determine the scale of the transformation needed to obtain the desired number of transformants

Figure 9.4 Flowchart of Step 1.

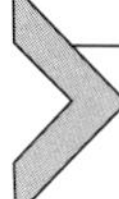

6. STEP 2 DETERMINE 3-AT CONCENTRATION TO BE USED IN SELECTION

6.1. Overview

3-AT is a competitive inhibitor of the *HIS3* gene product and is typically added to select for strong interactions. In this step, one can determine how much 3-AT to include in the selection, such that a balance between suppressing background activation and permitting growth of 'real' positives can be achieved.

6.2. Duration

4–6 days

2.1 Prepare a series of selective medium plates that lack histidine (SD–Leu–Ura–His) and contain increasing amounts of 3-AT (e.g., 0, 0.5, 1, 2, 3, and 5 mM). Also prepare selective medium plates containing histidine (SD–Leu–Ura). The SD–Leu–Ura–His plates are used to assess the level of background activation by the protein in the absence of an RNA. The SD–Leu–Ura plates serve as a control for the transformation.

2.2 Transform cells from a single yeast colony carrying the AD–protein fusion plasmid with an 'empty' hybrid RNA plasmid. Plate

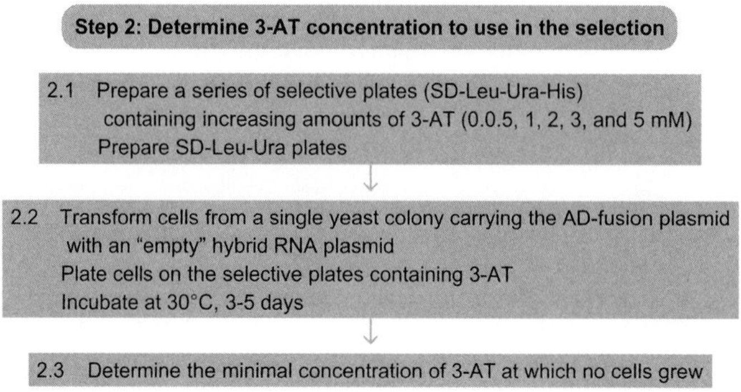

Step 2: Determine 3-AT concentration to use in the selection

2.1 Prepare a series of selective plates (SD-Leu-Ura-His) containing increasing amounts of 3-AT (0.0.5, 1, 2, 3, and 5 mM)
Prepare SD-Leu-Ura plates

↓

2.2 Transform cells from a single yeast colony carrying the AD-fusion plasmid with an "empty" hybrid RNA plasmid
Plate cells on the selective plates containing 3-AT
Incubate at 30°C, 3-5 days

↓

2.3 Determine the minimal concentration of 3-AT at which no cells grew

Figure 9.5 Flowchart of Step 2.

transformants on the prepared plates from Step 2.1. Incubate at 30 °C for 3–5 days.

2.3 Determine the minimal concentration of 3-AT that resulted in no growth of yeast transformants. That would be a suitable starting concentration for the selection experiment.

6.3. Tip

3-AT is light sensitive. Keep plates in the dark when possible.

6.4. Tip

If colonies are observed even at high concentrations of 3-AT, the protein likely autoactivates the reporter even in the absence of an interacting RNA and is not suitable for the selection experiment.

6.5. Tip

The addition of 3-AT to the medium decreases the number of false positives by demanding a more stringent interaction between the RNA and protein. However, 3-AT concentrations that are too high can result in the loss of colonies containing legitimate RNA–protein interactions.

See Fig. 9.5 for the flowchart of Step 2.

7. STEP 3 INTRODUCE THE HYBRID RNA LIBRARY

7.1. Overview

The hybrid RNA library is introduced into YBZ-1 yeast that has been transformed with the AD–protein fusion plasmid. Transformants are selected on SD–Leu–Ura–His and containing a predetermined amount of 3-AT from Step 2.

7.2. Duration

7–10 days

3.1 Transform cells from a single yeast colony carrying the AD–fusion plasmid with the RNA library using the LiAc method (Gietz and Woods, 2002, see Chemical Transformation of Yeast). Based on the expected transformation efficiency from Step 1 and the number of transformants desired, the scale of transformation should be adjusted accordingly.

3.2 Calculate the total number of transformants. Save 50 μl of transformed cells from Step 3.1 and make serial dilutions in sterile water (10^{-1}, 10^{-2}, 10^{-3}, and 10^{-4}). Plate 100 μl per dilution on SD–Leu–Ura medium and incubate at 30 °C for 3–4 days. Once colonies are visible, count the number of colonies and calculate the total number of transformants. The confidence level at which the entire library is screened can be calculated using the formula in Step 1.3.

3.3 Plate the remaining transformants on SD–Leu–Ura–His medium that contains a predetermined amount of 3-AT (from Step 2). Incubate at 30 °C for 7 days (or more, if necessary) until colonies start to form.

3.4 Pick the colonies from Step 3.3 and patch on fresh SD–Leu–Ura plates to maintain selection of both the RNA and AD–protein fusion plasmids. Typically, 50 positives can be patched onto a 100-mm diameter Petri dish.

See Fig. 9.6 for the flowchart of Step 3.

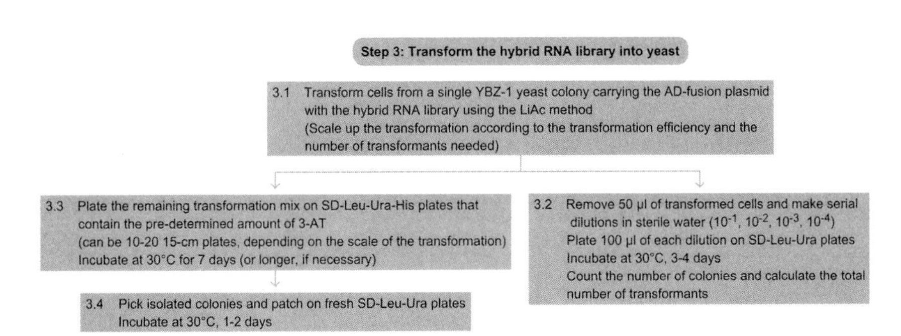

Figure 9.6 Flowchart of Step 3.

8. STEP 4 ASSAY β-GALACTOSIDASE ACTIVITY

8.1. Overview

As a secondary selection, the expression of *lacZ* is determined by performing a qualitative β-galactosidase activity. This step ensures that activation of *HIS3* is not spurious.

8.2. Duration

1–2 days

4.1 Place a nitrocellulose filter on a plate containing selective medium (SD–Leu–Ura) and patch transformants directly on the filter. Incubate cells at 30 °C for 1–2 days.

4.2 Lyse cells by immersing filter in liquid nitrogen for 5–20 s and thawing on the benchtop for 1 min.

4.3 Repeat freeze/thaw cycle twice.

4.4 Place 3MM Whatman paper in a Petri dish and saturate with 5 ml Z-buffer. Add 75 μl of 20 mg ml^{-1} X-Gal solution. Discard excess buffer in Petri dish.

4.5 Overlay the nitrocellulose filter containing cells onto saturated the Whatman paper, with the cells side up. Seal the dish with parafilm.

4.6 Incubate, colony side up, 30 min to overnight at 30 °C. Examine the filters regularly. A strong interaction (such as that between IRE and IRP) should turn blue within 30 min.

8.3. Caution

After freezing and thawing, the nitrocellulose filter is brittle and may crack easily.

8.4. Tip

X-Gal should be added fresh. Keep stock solution in the dark as it is light-sensitive.

8.5. Tip

Ensure good contact between the nitrocellulose filter and the Whatman paper.

8.6. Tip

With protracted incubation, weak interactions eventually yield a blue color. For this reason, it is important to examine the filters periodically to determine how long it takes

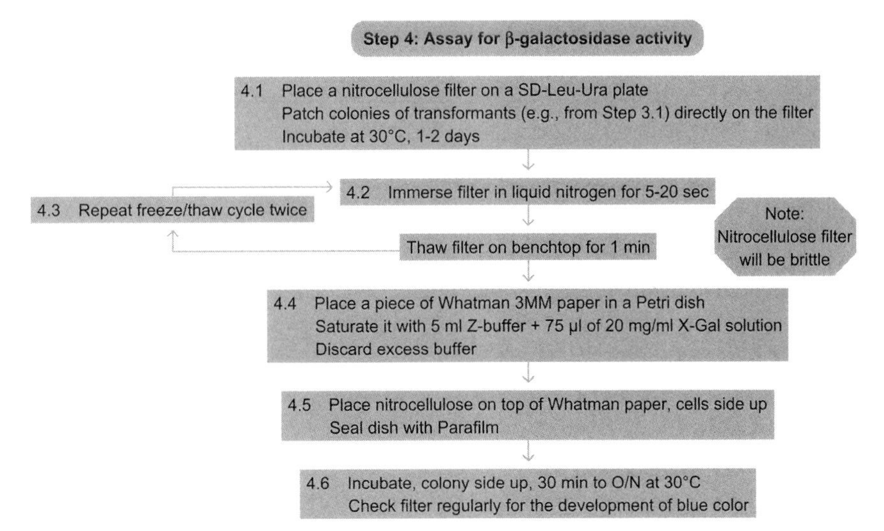

Figure 9.7 Flowchart of Step 4.

for the color to develop. It is also helpful to include a positive control (e.g., IRE and IRP interaction) to ensure that the assay works.

See Fig. 9.7 for the flowchart of Step 4.

9. STEP 5 CURE THE RNA PLASMID AND TEST POSITIVES FOR PROTEIN DEPENDENCE

9.1. Overview

One class of false positives that must be eliminated consists of hybrid RNAs that activate transcription of the reporter genes on their own, without interacting with the AD–protein fusion. This step classifies RNAs as protein-dependent, and thus worthy of additional analysis, or protein-independent, and thus of no further interest in this context. First, cells will be cured of the AD–protein fusion plasmid. Expression of *lacZ* is then monitored. Candidates that fail to activate the reporter genes are further analyzed.

9.2. Duration

4–5 days

5.1 Cure the AD–protein fusion plasmid by growing a positive transformant overnight at 30 °C in 5-ml YPAD medium and then plating for single colonies on SD–Ura to select for the RNA plasmid. Incubate the plates at 30 °C for 1–2 days.

Step 5: Cure the RNA plasmid and test positives for protein dependence

5.1 Grow a positive transformant overnight at 30°C in 5 ml YPAD medium
Plate for single colonies on SD-Ura plates (select for RNA plasmid)
Incubate at 30°C, 1-2 days

↓

5.2 Replica plate the cells onto SD-Leu plates to determine which colonies lack the AD-protein fusion plasmid
Incubate at 30°C, 1-2 days

↓

5.3 Assay yeast cells cured of the AD-protein fusion plasmid for β-galactosidase activity (as in step 4)

↓

5.4 Select candidates that fail to activate *lacZ* for further analysis
Discard colonies that turn blue (false positives since they activate *lacZ* in the absence of the AD-protein fusion plasmid)

Figure 9.8 Flowchart of Step 5.

5.2 Replica-plate the cells from Step 5.1 onto SD–Leu to determine which colonies lack the AD–protein fusion plasmid. Incubate the plates at 30 °C for 1–2 days. Colonies from the SD–Ura plates that did not grow on SD–Leu plates have been cured of the AD–protein fusion plasmid.
5.3 Assay cells that have been cured of the AD–protein fusion plasmid for β-galactosidase activity as in Step 4.
5.4 Select the candidates that fail to activate *lacZ* for further analysis. Discard false positives that activate *lacZ* (blue color) in the absence of the AD–protein fusion plasmid.

See Fig. 9.8 for the flowchart of Step 5.

10. STEP 6 ISOLATE PLASMIDS FOR AUTOACTIVATION TEST AND SEQUENCING

10.1. Overview

Once candidates are identified, the RNA plasmid must be recovered from yeast and introduced into *E. coli* (see Transformation of Chemically Competent *E. coli* or Transformation of *E. coli* via electroporation). The plasmids can then be amplified in *E. coli* for use in future applications. Hybrid RNA plasmids must be transformed back into a yeast strain containing an 'empty' AD plasmid to test for protein-independent activation (see

Chemical Transformation of Yeast). Expression of either the *HIS3* or *lacZ* reporter gene is monitored.

10.2. Duration

3–5 days

6.1 Inoculate each positive clone (from patched positives in Step 3) in 5-ml SD–Ura medium and grow to saturation (24–48 h) at 30 °C in a rotary shaker (200 rpm). This step selects for the RNA plasmid.

6.2 Pellet 2-ml cells at 5000 × *g*. Resuspend pellet in 250 μl of cell resuspension buffer.

6.3 Spheroplast yeast with 5-μl zymolyase for 2 h at 37 °C.

6.4 Add 250-μl lysis buffer and invert to mix.

6.5 Incubate for 5 min at room temperature, followed by 5 min at 65 °C. Cool to room temperature.

6.6 Add 10 μl of alkaline protease solution and incubate for 10 min at room temperature.

6.7 Add 350 μl of lysis buffer and invert to mix.

6.8 Centrifuge at > 16 000 × *g* for 10 min.

6.9 Add lysate to the binding column. Centrifuge at > 16 000 × *g* for 1 min.

6.10 Wash the column with a 700-μl wash buffer. Centrifuge at > 16 000 × *g* for 1 min. Discard the flow-through.

6.11 Wash the column with a 500-μl wash buffer. Centrifuge at > 16 000 × *g* for 1 min. Discard the flow-through.

6.12 Centrifuge at > 16 000 × *g* for 2 min to remove the residual buffer.

6.13 Elute the plasmid DNA in 100-μl water.

6.14 Transform the plasmid DNA into *E. coli* using conventional methods. Plate transformants on medium containing the appropriate antibiotic to select for the RNA plasmid.

6.15 Perform colony PCR to identify clones carrying the RNA plasmid (see Colony PCR). Isolate plasmids from *E. coli* using conventional methods.

6.16 Transform each RNA plasmid back into a yeast strain containing an 'empty' AD–protein fusion plasmid.

6.17 Assay transformants for the expression of the *lacZ* reporter, as described in Step 4.

6.18 Discard false positives that activate *lacZ* in the absence of your specific RNA-binding protein. Identify the remaining RNAs by sequencing the plasmids isolated from *E. coli*.

10.3. Tip

This step is a modified version of the Promega Wizard Plus SV Miniprep kit.

10.4. Tip

Since either RNA or cDNA plasmids can be present in E. coli, *colonies can be screened for the plasmid of interest by colony PCR. If different antibiotic resistance markers are present on each plasmid, most colonies should contain the RNA plasmid.*

10.5. Tip

Instead of assaying for expression of the lacZ reporter, HIS3 expression can be monitored as described in Dissecting a known RNA-protein interaction using a yeast three-hybrid system.

10.6. Tip

Compare the sequences to common sequence databases to identify RNA targets (e.g., BLAST). If a consensus binding sequence is desired, the RNA sequences can be analyzed using motif discovery software (e.g., MEME).

See Fig. 9.9 for the flowchart of Step 6.

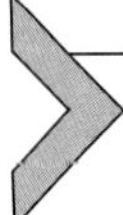

11. STEP 7 DETERMINE BINDING SPECIFICITY USING MUTANT AND CONTROL PROTEINS

11.1. Overview

Clones that require the AD–protein fusion plasmid to activate the reporter genes can then be analyzed to determine whether they interact sequence specifically with the protein.

11.2. Duration

3–4 days

7.1 Transform YBZ-1 yeast with control activation domain plasmids expressing a mutant protein that is deficient in RNA-binding.

7.2 Transform the candidate RNA plasmids into a yeast strain containing either the wild-type or mutant AD–protein fusion plasmids.

7.3 Assay transformants for the expression of the *lacZ* reporter qualitatively, as described in Step 4, to determine sequence specificity of the positive clones.

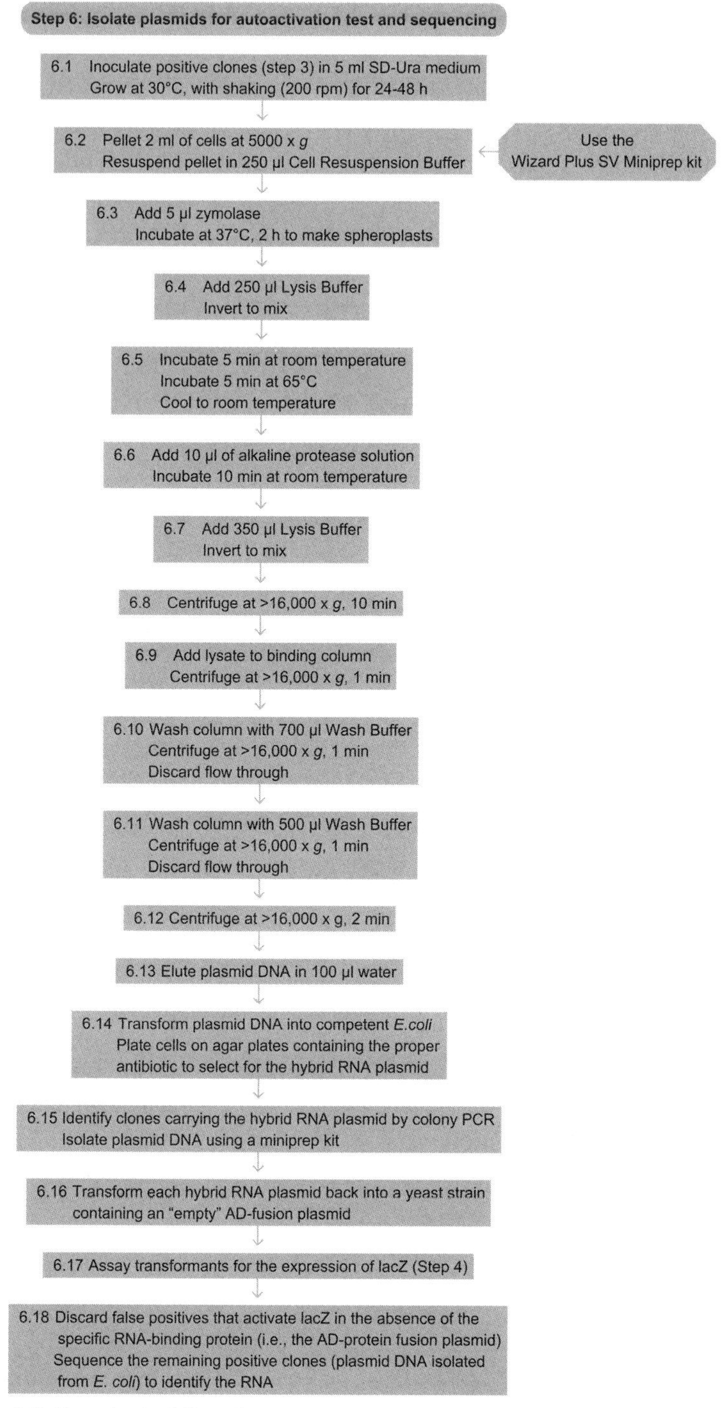

Figure 9.9 Flowchart of Step 6.

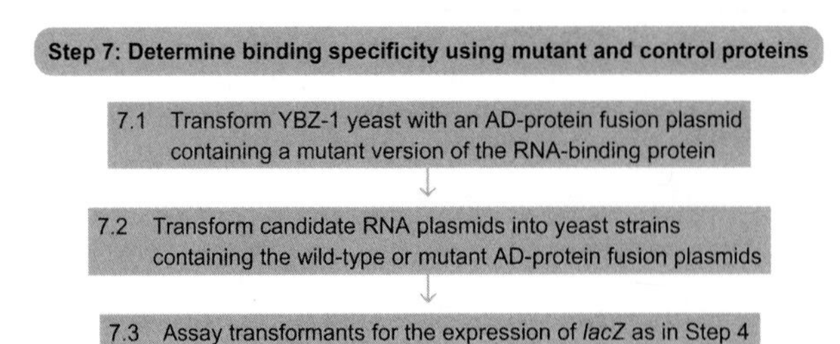

Figure 9.10 Flowchart of Step 7

11.3. Tip

The more subtle the mutation used, the better. If no subtle mutations are available, rudimentary analyses such as using deletions or an unrelated protein can be helpful.

11.4. Tip

If a quantitative comparison between protein and various RNAs is desired, a quantitative β-galactosidase assay is provided in Dissecting a known RNA-protein interaction using a yeast three-hybrid system.

See Fig. 9.10 for the flowchart of Step 7.

12. STEP 8 FUNCTIONAL TESTS OR ADDITIONAL SCREENS

12.1. Overview

Almost invariably, additional steps will be needed to identify those positives that are biologically meaningful. Each screen is unique. The interactions being analyzed and the organisms being studied will determine what additional steps need to be taken to determine the biological relevance of each interaction. It is not surprising, given that the assay is performed outside of most biological contexts, that some specific, high-affinity interactions may not be relevant to the biology of the system being studied.

REFERENCES

Referenced Literature

Gietz, R. D., & Woods, R. A. (2002). Transformation of yeast by lithium acetate/single-stranded carrier DNA/polyethylene glycol method. *Methods in Enzymology, 350*, 87–96.

Hook, B., Bernstein, D., Zhang, B., & Wickens, M. (2005). RNA-protein interactions in the yeast three-hybrid system: Affinity, sensitivity, and enhanced library screening. *RNA, 11*, 227–233.

Koh, Y. Y., Opperman, L., Stumpf, C., Mandan, A., Keles, S., & Wickens, M. (2009). A single C. elegans PUF protein binds RNA in multiple modes. *RNA, 15*, 1090–1099.

Sengupta, D. J., Wickens, M., & Fields, S. (1999). Identification of RNAs that bind to a specific protein using the yeast three-hybrid system. *RNA*, *5*, 596–601.
Stumpf, C. R., Kimble, J., & Wickens, M. (2008). A Caenorhabditis elegans PUF protein family with distinct RNA binding specificity. *RNA*, *14*, 1550–1557.
Stumpf, C. R., Opperman, L., & Wickens, M. (2008). Chapter 14. Analysis of RNA-protein interactions using a yeast three-hybrid system. *Methods in Enzymology*, *449*, 295–315.
Zhang, B., Gallegos, M., Puoti, A., et al. (1997). A conserved RNA-binding protein that regulates sexual fates in the C. elegans hermaphrodite germ line. *Nature*, *390*, 477–484.

SOURCE REFERENCES

Bernstein, D. S., Buter, N., Stumpf, C., & Wickens, M. (2002). Analyzing mRNA-protein complexes using a yeast three-hybrid system. *Methods*, *26*, 123–141.
Stumpf, C. R., Opperman, L., & Wickens, M. (2008). Chapter 14. Analysis of RNA-protein interactions using a yeast three-hybrid system. *Methods in Enzymology*, *449*, 295–315.

Referenced Protocols in Methods Navigator

UV crosslinking of interacting RNA and protein in cultured cells.
PAR-CLIP (Photoactivatable Ribonucleoside-Enhanced Crosslinking and Immunoprecipitation): a step-by-step protocol to the transcriptome-wide identification of binding sites of RNA-binding proteins.
Chemical Transformation of Yeast.
Saccharomyces cerevisiae Growth Media.
Transformation of Chemically Competent *E. coli*.
Transformation of *E. coli* via electroporation.
Colony PCR.
Dissecting a known RNA-protein interaction using a yeast three-hybrid system.

CHAPTER TEN

Dissecting a Known RNA–Protein Interaction using a Yeast Three-Hybrid System

Yvonne Y. Koh*, **Marvin Wickens**[†,1]
*Microbiology Doctoral Training Program, University of Wisconsin – Madison, Madison, WI, USA
†Department of Biochemistry, University of Wisconsin – Madison, Madison, WI, USA
[1]Corresponding author: e-mail address: wickens@biochem.wisc.edu

Contents

Methods in Enzymology, Volume 539
ISSN 0076-6879
http://dx.doi.org/10.1016/B978-0-12-420120-0.00010-4

Abstract

The yeast three-hybrid system has been applied to known protein–RNA interactions for a variety of purposes. For instance, protein and RNA mutants with altered or relaxed binding specificities can be identified. Mutant RNAs can also be analyzed to better understand RNA-binding specificity of a specific protein. Furthermore, this system complements other biochemical techniques, for example, SELEX, co-immunoprecipitation and cross-linking experiments (see UV crosslinking of interacting RNA and protein in cultured cells and PAR-CLIP (Photoactivatable Ribonucleoside-Enhanced Crosslinking and Immunoprecipitation): a step-by-step protocol to the transcriptome-wide identification of binding sites of RNA-binding proteins).

1. THEORY

The general strategy of the three-hybrid system is shown in Fig. 10.1. DNA-binding sites are placed upstream of a reporter gene, which has been

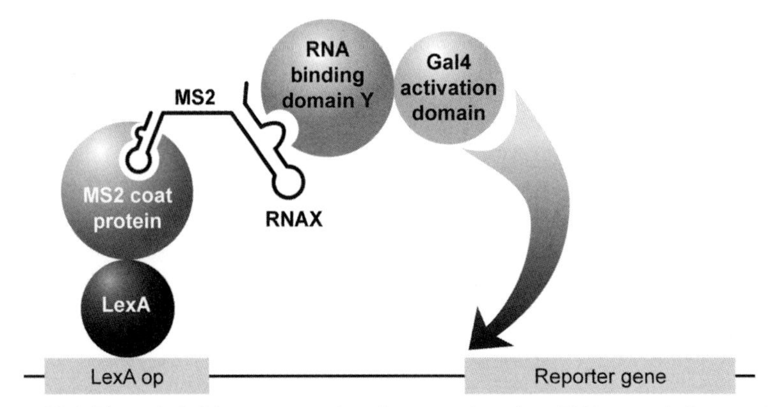

Figure 10.1 Three-hybrid system used to detect and analyze RNA–protein interactions. The diagram depicts the general strategy of the three-hybrid system. Specific protein and RNA components that are typically used are represented. For the sake of simplicity, the following points are not illustrated. In the YBZ-1 strain, both *lacZ* and *HIS3* reporter genes are present under the control of lexA operators (eight in the *lacZ* promoter and four in the *HIS3* promoter). The LexA protein binds as a dimer. The hybrid RNA contains two MS2 stem-loops and the MS2 coat protein binds one stem-loop as a dimer.

integrated into the yeast genome. The first hybrid protein consists of a DNA-binding domain linked to an RNA-binding domain. The RNA-binding domain interacts with its RNA-binding site in a bifunctional ('hybrid') RNA molecule. The other part of the RNA molecule interacts with a second hybrid protein consisting of another RNA-binding domain linked to a transcription activation domain. When this tripartite complex forms at the promoter, the reporter gene is turned on. Reporter expression can be detected by phenotype or simple biochemical assays. The specific molecules used most commonly for three-hybrid analysis are depicted in Fig. 10.1. The DNA-binding site consists of a 17-nucleotide recognition site for the *Escherichia coli* LexA protein and is present in multiple copies upstream of both the *HIS3* and the *lacZ* genes. The first hybrid protein consists of LexA fused to bacteriophage MS2 coat protein, a small polypeptide that binds to a short stem-loop sequence present in the bacteriophage RNA genome. The hybrid RNA (depicted in more detail in Fig. 10.2) consists of two MS2 coat protein-binding

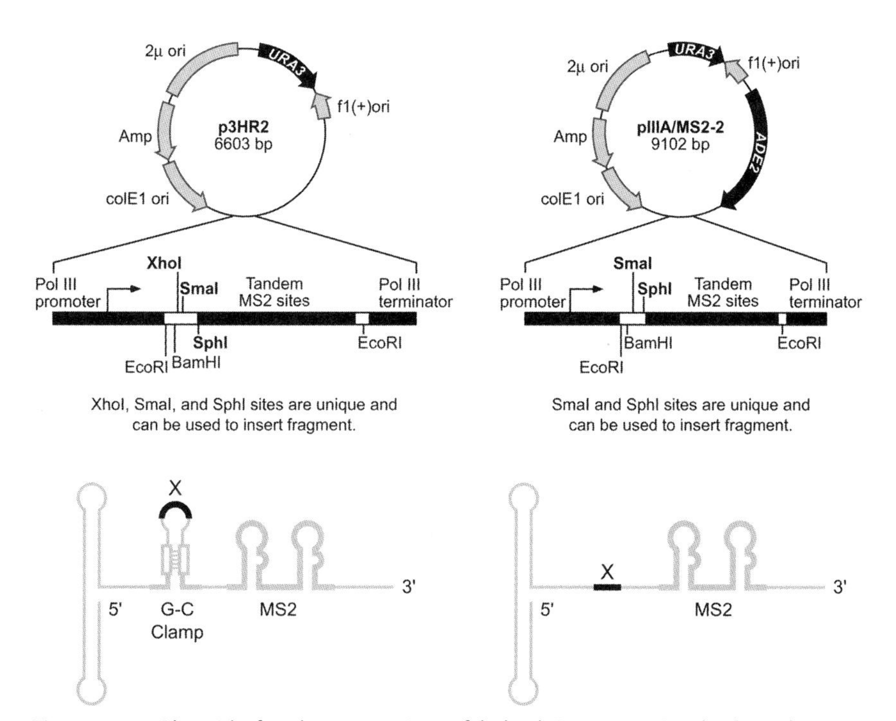

Figure 10.2 Plasmids for the expression of hybrid RNAs. p3HR2 displays the RNA sequence of interest in the loop of a G-C clamp (Stumpf et al., 2008), whereas pIIIA/MS2-2 does not (Zhang et al., 1997). The sequence of interest can be inserted into the unique restriction sites.

sites linked to the RNA sequence of interest, X. The second hybrid protein consists of the transcription activation domain of the yeast Gal4 transcription factor linked to an RNA-binding protein, Y.

Here, we consider a case in which RNA X and protein Y are known, and their interaction is assayed *in vivo*. This protocol is intended to analyze mutant RNAs and proteins. The output of the system – *HIS3* and *lacZ* activation – has been examined as a function of *in vitro* affinity using a specific RNA–protein interaction (Hook et al., 2005). When the protein FBF-1 was tested with various RNAs, K_d was linearly related to the log of β-galactosidase activity over the range measured, from 1 to 100 nM.

2. EQUIPMENT

Centrifuge
Spectrophotometer
Luminometer (e.g., Turner 20/20n Single Tube Luminometer, Promega)
30 °C incubator-shaker
Vortex mixer
Micropipettors
Micropipettor tips
1-ml cuvettes
Luminometer tubes
Glass tubes (18 × 150 mm) with caps
Petri dishes (100 mm)
Nitrocellulose filter (Petri dish-sized)
Whatman 3MM chromatography paper (Petri dish-sized)
Parafilm
Sterile toothpicks
0.2-μm filters
1.5-ml microcentrifuge tubes

3. MATERIALS

Saccharomyces cerevisiae strain YBZ-1: *MATa, ura3-52, leu2-3, -112, his3-200, trp1-1, ade2, LYS2::(LexAop)-HIS3, URA3::(lexAop)-lacZ, and LexA-MS2 MS2 coat (N55K)*
Hybrid RNA plasmid (e.g., pIIIA/MS2-2 or p3HR2)
Activation domain fusion plasmid (e.g., pACTII or pGADT7)

Synthetic dextrose (SD) dropout media
Sodium phosphate dibasic (Na_2HPO_4)
Sodium phosphate monobasic (NaH_2PO_4)
Potassium chloride (KCl)
Magnesium sulfate ($MgSO_4$)
2-Mercaptoethanol
5-Bromo-4-chloro-3-indolyl-β-D-galactopyranoside (X-Gal)
Dimethylformamide
3-Aminotriazole
Beta-Glo® reagent (Promega)
Liquid nitrogen

3.1. Solutions & buffers

Step 1A Z-buffer

Component	Final concentration	Stock	Amount
Na_2HPO_4	60 mM	1 M	60 ml
NaH_2PO_4	40 mM	1 M	40 ml
KCl	10 mM	1 M	10 ml
$MgSO_4$	1 mM	1 M	1 ml
2-Mercaptoethanol	50 mM	14.2 M	Add fresh

Add to 850 ml water. Adjust the pH to 7.0, if necessary. Add water to 1 l and filter sterilize.

X-Gal solution (20 mg ml^{-1})

Dissolve 0.4 g X-Gal in 20 ml dimethylformamide (in a fume hood). Aliquot into tubes, wrap in foil and store at −20 °C.

Step 1C 3-Aminotriazole (1 M)

Dissolve 4.21 g 3-AT in 50 ml water. Pass through a 0.2-μm filter to sterilize. Aliquot into tubes, wrap in foil, and store at −50 °C.

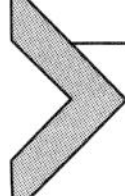

4. PROTOCOL

4.1. Duration

Preparation	About 5–9 days
Protocol	2–5 days

4.2. Preparation

The protein(s) of interest and desired RNA sequence(s) to be tested should be cloned into an activation domain fusion vector and hybrid RNA vector respectively (as a reference see Molecular Cloning). Here, we consider a case in which the activation domain vector carries a LEU2 marker and the RNA vector carries the URA3 marker.

Introduce both protein and RNA plasmids into YBZ-1 yeast using standard transformation protocols (Gietz and Woods, 2002, see also Chemical Transformation of Yeast). Transformants should be plated on SD–Leu–Ura medium (see *Saccharomyces cerevisiae* Growth Media).

4.3. Tip

Yeast can be transformed with both plasmids simultaneously or the fusion protein plasmid can be transformed first, followed by the RNA plasmid. Usually, transforming the plasmids sequentially yields more transformants. However, the fusion protein plasmid is occasionally toxic to the yeast, requiring a cotransformation of the plasmids.

4.4. Tip

For a detailed yeast transformation protocol, refer to the Gietz Lab webpage (http://home.cc.umanitoba.ca/~gietz/index.html).

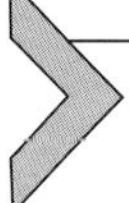

5. STEP 1A ASSAYING INTERACTIONS: QUALITATIVE FILTER ASSAY FOR β-GALACTOSIDASE ACTIVITY

5.1. Overview

The 'strength' of an RNA–protein interaction is gauged by assaying the activity of a reporter gene, in this case, *lacZ*. β-galactosidase activity can be assayed by measuring the conversion of a lactose analog to a chromogenic or luminescent product. This is a qualitative assay for β-galactosidase activity, carried out on yeast grown on a nitrocellulose filter.

5.2. Duration

2 days

1A.1 Place a nitrocellulose filter on a plate containing selective medium (SD–Leu–Ura) and streak transformants directly on the filter. Incubate cells at 30 °C for 1–2 days.

1A.2 Lyse cells by immersing the filter in liquid nitrogen for 5–20 s and thawing on the benchtop for 1 min.

1A.3 Repeat the freeze/thaw cycle twice.

1A.4 Place Whatman 3MM chromatography paper in a Petri dish and saturate it with 5 ml Z-buffer. Add 75 μl of 20 mg ml^{-1} X-Gal solution. Remove excess buffer in the Petri dish.

1A.5 Place the nitrocellulose filter containing the cells onto the saturated Whatman paper, with the cells side up. Seal the dish with Parafilm.

1A.6 Incubate the filter, with the colony side up, for 30 min to overnight at 30 °C. Examine the filters regularly. A strong interaction (such as that between IRE and IRP) should turn blue within 30 min.

5.3. Tip

Pick multiple average-sized colonies for each inoculation. While this is not ideal microbiological practice, it has worked reproducibly.

5.4. Caution

After freezing and thawing, the nitrocellulose filter is brittle and may crack easily.

5.5. Tip

X-Gal should be added to the Z-buffer fresh.

5.6. Tip

Ensure good contact between the nitrocellulose filter and Whatman paper.

5.7. Tip

With protracted incubation, weak interactions eventually yield a blue color. For this reason, it is important to examine the filters periodically to determine how long it takes for the color to develop. It is also helpful to include a positive control (e.g., IRE and IRP interaction) to ensure that the assay works.

See Fig. 10.3 for the flowchart of Step 1A.

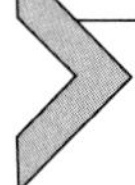

6. STEP 1B ASSAYING INTERACTIONS: QUANTITATIVE SOLUTION ASSAY FOR β-GALACTOSIDASE ACTIVITY

6.1. Overview

The ‘strength’ of an RNA–protein interaction is gauged by assaying the activity of a reporter gene, in this case, *lacZ*. β-galactosidase activity can be assayed by measuring the conversion of a lactose analog to a chromogenic or luminescent product. This is a quantitative liquid β-galactosidase assay

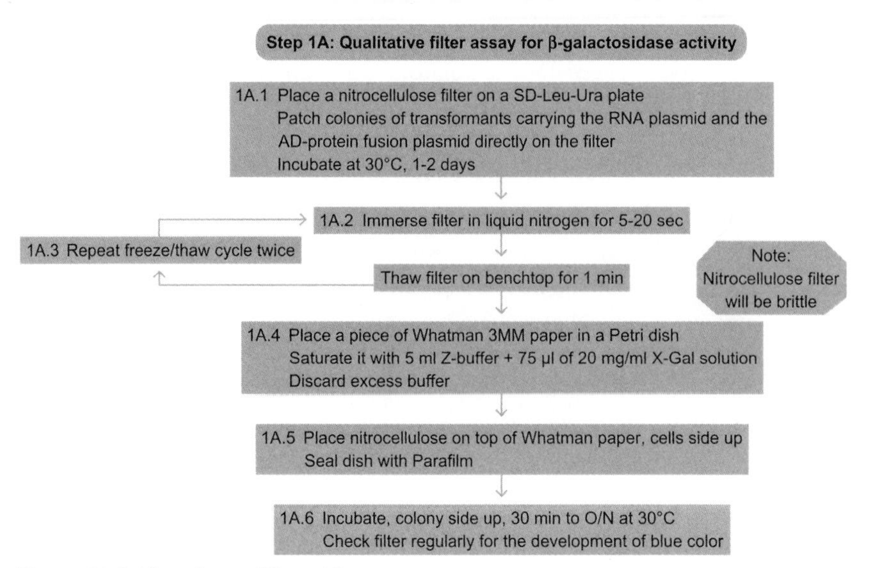

Figure 10.3 Flowchart of Step 1A.

using the enzyme-coupled luminescent substrate, Beta-Glo®. This simple yet sensitive assay uses a luminometer to measure the output from the lacZ gene.

6.2. Duration

2 days

1B.1 Inoculate a 4 ml culture for each strain in selective media (SD–Leu–Ura) in triplicate. Grow the cultures to saturation on a rotary shaker (200 rpm) at 30 °C (1.5–2 days).

1B.2 Dilute the cultures 1:40 (100 µl in 4 ml fresh media) and let them grow for 2–2.5 h. Check 1 ml of the culture to ensure an OD_{660} of 0.1–0.2.

1B.3 Thaw the Beta-Glo® reagent and let it warm to room temperature.

1B.4 Dispense 50 µl aliquots of Beta-Glo® reagent into enough microcentrifuge tubes for the assay. Add 50 µl of exponentially growing cells to the tubes. Vortex each tube for 10 s to ensure complete lysis.

1B.5 Incubate for 1 h at room temperature in the dark.

1B.6 Transfer 10 µl of the lysate to an appropriate luminometer tube.

1B.7 Integrate luminiscent signal for 1 s.

1B.8 Normalize signal to cell number (or 0.1 OD_{660}) to yield an activity/cell (or activity/0.1 OD_{660}) value.

6.3. Tip

Pick multiple average-sized colonies for each inoculation. While this is not ideal microbiological practice, it has worked reproducibly.

6.4. Tip

Yeast cells tend to settle quickly. Vortex the cultures to ensure that they are mixed well before reading OD_{660}.

6.5. Caution

The Beta-Glo® reagent is light sensitive. Thaw it in the dark.

6.6. Tip

Luminescence is stable from 30 min up to 4 h after adding the reagent.

6.7. Tip

Numerous variations on the liquid β-galactosidase assay exist. The protocol described here uses the enzyme-coupled luminescent substrate Beta-Glo® (Promega Corporation). This simple yet sensitive assay uses a luminometer to measure the output from the lacZ gene. The protocol described here uses a Turner 20/20n luminometer (Promega Corporation). Certain details, such as sample volumes, will vary depending on the instrument used.

6.8. Tip

This protocol can be modified to read a large number of assays by using a microplate luminometer.

See Fig. 10.4 for the flowchart of Step 1B.

7. STEP 1C ASSAYING INTERACTIONS: 3-AMINOTRIAZOLE (3-AT) RESISTANCE ASSAY

7.1. Overview

The 'strength' of an RNA–protein interaction is gauged by assaying the activity of a reporter gene, in this case, *HIS3*. This step determines the level of resistance to 3-aminotriazole (3-AT), which monitors *HIS3* activity.

7.2. Duration

3–5 days

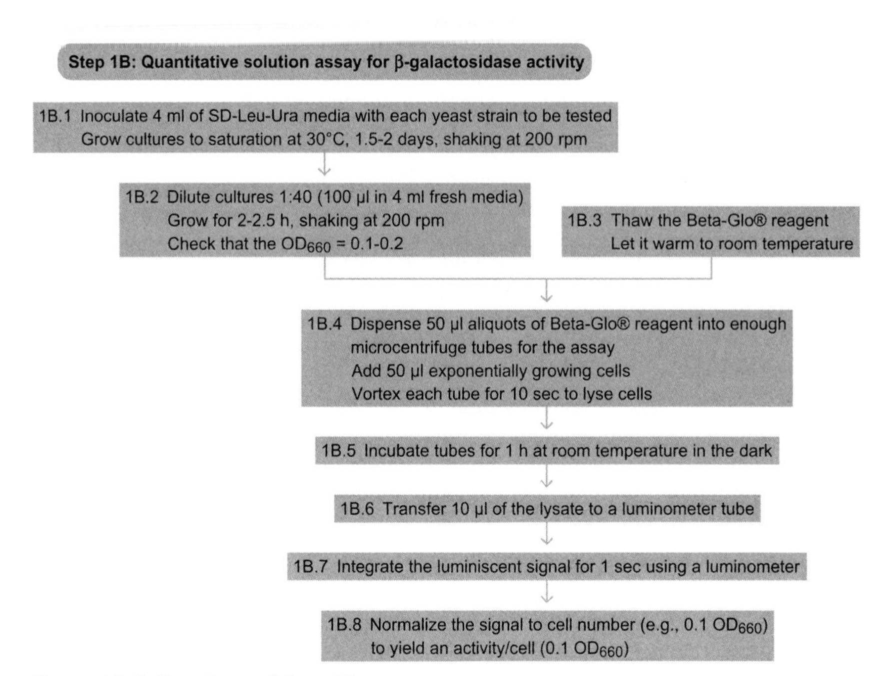

Figure 10.4 Flowchart of Step 1B.

1C.1 Prepare a series of selective medium plates that lack histidine (SD–Leu–Ura–His) and contain increasing amounts of 3-AT (e.g., 0, 0.5, 1, 5, 10, 15, 20 mM). Also prepare selective medium plates containing histidine (SD–Leu–Ura) to serve as a control for the transformation.

1C.2 Pick multiple average-sized transformants and streak for single colonies on prepared plates.

1C.3 Incubate at 30 °C for 3–5 days. Determine the minimal concentration of 3-AT needed to inhibit growth of the yeast. Growth should be assessed by the presence of individual colonies and not the smear of yeast from the initial streak.

7.3. Tip

3-AT is light sensitive. Keep the plates in the dark when possible.

7.4. Tip

While this is not ideal microbiological practice, it has worked reproducibly.

Step 1C: 3-Aminotriazole (3-AT) resistance assay

1C.1 Prepare a series of SD-Leu-Ura-His plates containing increasing concentrations of 3-AT (0, 0.5, 1, 5, 10, 15, 20 mM)
Prepare SD-Leu-Ura plates

↓

1C.2 Pick multiple average-sized transformants and streak for single colonies on the selective plates

↓

1C.3 Incubate at 30°C, 3-5 days
Determine the minimal concentration of 3-AT needed to inhibit growth of the yeast
Assess growth by the presence of individual colonies (not the smear from the initial streak)

Figure 10.5 Flowchart of Step 1C.

7.5. Caution

It is common to observe a few large colonies on high concentrations of 3-AT.

See Fig. 10.5 for the flowchart of Step 1C.

REFERENCES

Referenced Literature

Gietz, R. D., & Woods, R. A. (2002). Transformation of yeast by lithium acetate/single-stranded carrier DNA/polyethylene glycol method. *Methods in Enzymology*, *350*, 87–96.

Hook, B., Bernstein, D., Zhang, B., & Wickens, M. (2005). RNA-protein interactions in the yeast three-hybrid system: Affinity, sensitivity, and enhanced library screening. *RNA*, *11*, 227–233.

Stumpf, C. R., Opperman, L., & Wickens, M. (2008). Chapter 14. Analysis of RNA-protein interactions using a yeast three-hybrid system. *Methods in Enzymology*, *449*, 295–315.

Zhang, B., Gallegos, M., Puoti, A., et al. (1997). A conserved RNA-binding protein that regulates sexual fates in the C. elegans hermaphrodite germ line. *Nature*, *390*, 477–484.

SOURCE REFERENCES

Bernstein, D. S., Buter, N., Stumpf, C., & Wickens, M. (2002). Analyzing mRNA-protein complexes using a yeast three-hybrid system. *Methods*, *26*, 123–141.

Stumpf, C. R., Opperman, L., & Wickens, M. (2008). Chapter 14. Analysis of RNA-protein interactions using a yeast three-hybrid system. *Methods in Enzymology*, *449*, 295–315.

Referenced Protocols in Methods Navigator

UV crosslinking of interacting RNA and protein in cultured cells.

PAR-CLIP (Photoactivatable Ribonucleoside-Enhanced Crosslinking and Immunoprecipitation): a step-by-step protocol to the transcriptome-wide identification of binding sites of RNA-binding proteins.

Molecular Cloning.

Chemical Transformation of Yeast.

Saccharomyces cerevisiae Growth Media.

CHAPTER ELEVEN

Identifying Proteins that Bind a Known RNA Sequence Using the Yeast Three-Hybrid System

Yvonne Y. Koh*, Marvin Wickens†,1

*Microbiology Doctoral Training Program, University of Wisconsin – Madison, Madison, WI, USA
†Department of Biochemistry, University of Wisconsin – Madison, Madison, WI, USA
[1]Corresponding author: e-mail address: wickens@biochem.wisc.edu

Contents

Methods in Enzymology, Volume 539
ISSN 0076-6879
http://dx.doi.org/10.1016/B978-0-12-420120-0.00011-6

Abstract

The yeast three-hybrid system can be used to identify a protein partner of a known RNA sequence by screening a cDNA library fused to a transcription activation domain, with a hybrid RNA as 'bait.' Most commonly, such screens are performed to identify proteins that interact with a given RNA *in vivo*.

1. THEORY

The general strategy of the three-hybrid system is shown in Fig. 11.1. DNA-binding sites are placed upstream of a reporter gene, which has been integrated into the yeast genome. The first hybrid protein consists of a DNA-binding domain linked to an RNA-binding domain. The RNA-binding domain interacts with its RNA-binding site in a bifunctional

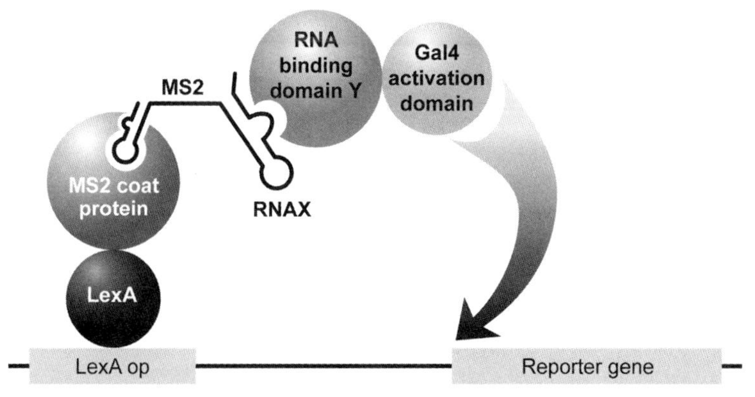

Figure 11.1 Three-hybrid system used to detect and analyze RNA–protein interactions. The diagram depicts the general strategy of the three-hybrid system. Specific protein and RNA components that are typically used are represented. For the sake of simplicity, the following points are not illustrated. In the YBZ-1 strain, both *lacZ* and *HIS3* reporter genes are present under the control of lexA operators (eight in the *lacZ* promoter and four in the *HIS3* promoter). The LexA protein binds as a dimer. The hybrid RNA contains two MS2 stem-loops and the MS2 coat protein binds one stem-loop as a dimer.

('hybrid') RNA molecule. The other part of the RNA molecule interacts with a second hybrid protein consisting of another RNA-binding domain linked to a transcription activation domain. When this tripartite complex forms at the promoter, the reporter gene is turned on. Reporter expression can be detected by phenotype or simple biochemical assays. The specific molecules used most commonly for three-hybrid analysis are depicted in Fig. 11.1. The DNA-binding site consists of a 17-nucleotide recognition site for the *Escherichia coli* LexA protein and is present in multiple copies upstream of both the *HIS3* and the *lacZ* genes. The first hybrid protein consists of LexA fused to bacteriophage MS2 coat protein, a small polypeptide that binds to a short stem-loop sequence present in the bacteriophage RNA genome. The hybrid RNA (depicted in more detail in Fig. 11.2) consists of two MS2 coat protein-binding sites linked to the RNA sequence of interest, X. The second hybrid protein consists of the transcription activation domain of the yeast Gal4 transcription factor linked to an RNA-binding protein, Y.

By introducing a cDNA library, cognate protein partners of a known RNA sequence can be identified. Secondary screens of the initial 'positives' are needed to winnow down candidates for further analysis and to eliminate false positives. Here, we consider a case in which the activation domain library carries a *LEU2* marker and the hybrid RNA vector (pIIIA) carries the *ADE2* and *URA3* markers. When the RNA interacts with the protein

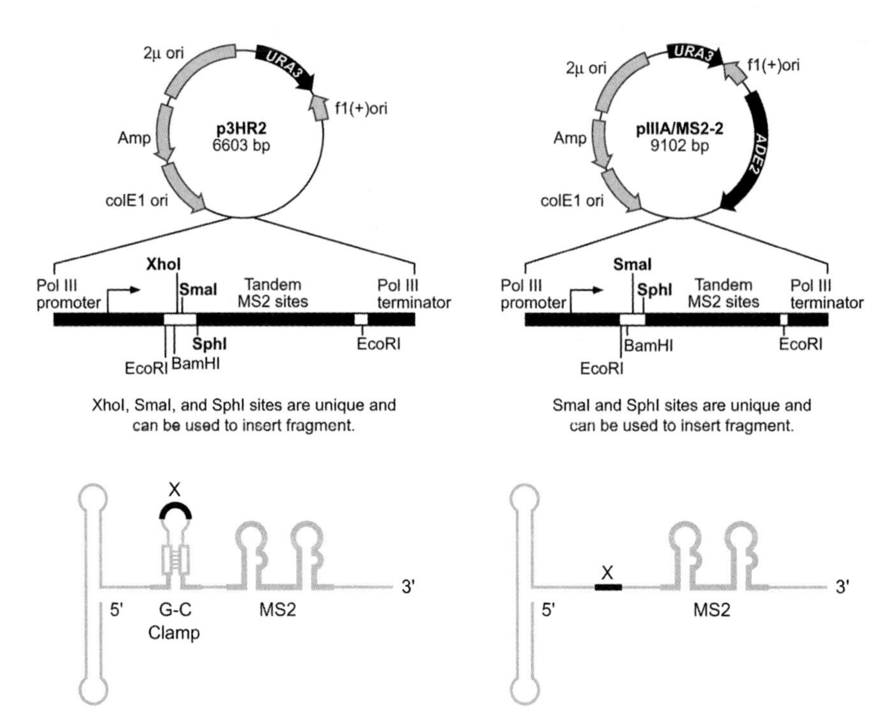

Figure 11.2 Plasmids for the expression of hybrid RNAs. p3HR2 displays the RNA sequence of interest in the loop of a G-C clamp (Stumpf et al., 2008), whereas pIIIA/MS2-2 does not (Zhang et al., 1997). The sequence of interest can be inserted into the unique restriction sites. pIIIA/MS2-2 carries an *ADE2* marker which can be used in Step 4 to eliminate false positives by screening colony color.

produced from a cDNA, the *HIS3* reporter gene is expressed and the yeast grow on media lacking histidine and/or containing 3-aminotriazole (a competitive inhibitor of the *HIS3* gene product). Additional screens are performed using the *lacZ* reporter and colorimetric assays.

2. EQUIPMENT

Centrifuge
30 °C incubator-shaker
Petri dishes
Micropipettors
Micropipettor tips
Nitrocellulose filter (Petri dish-sized)
Whatman 3MM chromatography paper (Petri dish-sized)
Sterile toothpicks
Parafilm

3. MATERIALS

cDNA–activation domain fusion library

Saccharomyces cerevisiae strain YBZ-1: *MATa, ura3-52, leu2-3, -112, his3-200, trp1-1, ade2, LYS2::(LexAop)-HIS3, URA3::(lexAop)-lacZ, and LexA-MS2 MS2 coat (N55K)*

Hybrid RNA plasmid (e.g., pIIIA/MS2-2)

Synthetic dextrose (SD) dropout media (liquid and plates)

3-Aminotriazole (3-AT)

5-Fluoroorotic acid (5-FOA)

Sodium phosphate dibasic (Na_2HPO_4)

Sodium phosphate monobasic (NaH_2PO_4)

Potassium chloride (KCl)

Magnesium sulfate ($MgSO_4$)

2-Mercaptoethanol

5-Bromo-4-chloro-3-indolyl-β-D-galactopyranoside (X-Gal)

Dimethylformamide

Zymolyase (G-Biosciences)

Promega Wizard Plus SV Miniprep kit

Liquid nitrogen

3.1. Solutions & buffers

Step 2 3-Aminotriazole (1 M)

Dissolve 4.21-g 3-AT in 50-ml water. Pass through a 0.2-μm filter to sterilize. Aliquot into tubes, wrap in foil, and store at −50 °C.

Step 5 Z-buffer

Component	Final concentration	Stock	Amount
Na_2HPO_4	60 mM	1 M	60 ml
NaH_2PO_4	40 mM	1 M	40 ml
KCl	10 mM	1 M	10 ml
$MgSO_4$	1 mM	1 M	1 ml
2-Mercaptoethanol	50 mM	14.2 M	Add fresh

Add to 850-ml water. Adjust pH to 7.0, if necessary. Add water to 1 l and filter-sterilize.

X-Gal solution (20 mg ml^{-1})
Dissolve 0.4-g X-Gal in 20-ml dimethylformamide (in a fume hood). Aliquot into tubes, wrap in foil, and store at −20 °C.

4. PROTOCOL

4.1. Duration

Preparation	1 week
Protocol	1–2 months

4.2. Preparation

Obtain appropriate activation domain library either from individual laboratories or commercial sources.

Transform the hybrid RNA plasmid into YBZ-1 yeast using the LiAc method (Gietz and Woods, 2002, see Chemical Transformation of Yeast).

Prepare sufficient plates listed in subsequent steps for the selection (see Saccharomyces cerevisiae Growth Media*).*

See Fig. 11.3 for the flowchart of the complete protocol.

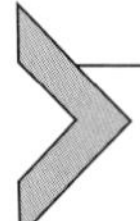

5. STEP 1 PILOT TRANSFORMATION TO DETERMINE EXPECTED TRANSFORMATION EFFICIENCY

5.1. Overview

Perform a pilot transformation to determine the expected transformation efficiency. This allows one to calculate how much to scale up the transformation in order to obtain the desired number of transformants.

5.2. Duration

3–4 days

1.1 Transform cells from a single YBZ-1 yeast colony carrying the RNA plasmid with the activation domain library using the LiAc method (Gietz and Woods, 2002, see Chemical Transformation of Yeast).

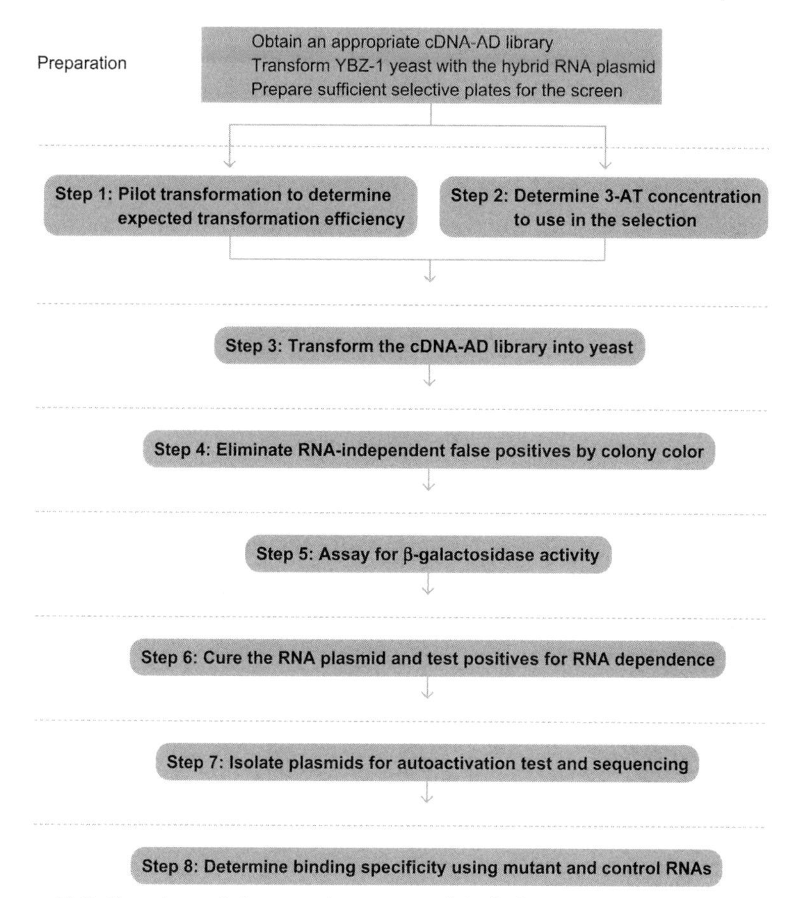

Figure 11.3 Flowchart of the complete protocol, including preparation.

Select transformants by plating on SD–Leu–Ura medium. Incubate the plates at 30 °C for 3–4 days.

1.2 Count the number of colonies from a single transformation ($\sim 10^8$ cells) to determine transformation efficiency (number of transformants/μg DNA/10^8 cells).

1.3 Calculate the number of transformants desired to obtain sufficient coverage of the library. For example, to be 95% confident that a library with 10^6 clones is entirely screened, one must search through 3×10^6 transformants as determined by the following equation:

$$N = \ln(1-p)/\ln(1-(1/u)),$$

where N, transformants; p, probability (% confidence); u, library size.

1.4 *Based on the transformation efficiency from Step 1.2, determine the scale of transformation required to obtain the desired number of transformants in Step 1.3.*

5.3. Tip

Steps 1 and 2 (pilot transformation and determination of 3-AT concentration to use) can be carried out simultaneously.

5.4. Tip

Yeast can be transformed with both plasmids simultaneously or sequentially. Usually, transformation of the plasmids sequentially yields more transformants. However, transformation a single plasmid is occasionally toxic to the yeast, requiring a cotransformation of both plasmids.

5.5. Tip

For starters, transform with 0.1–1.0-μg library plasmid DNA. The amount of DNA has to be optimized to obtain a high transformation efficiency.

5.6. Tip

For a detailed yeast transformation protocol, refer to the Gietz Lab webpage (http://home.cc.umanitoba.ca/~gietz/index.html).

See Fig. 11.4 for the flowchart of Step 1.

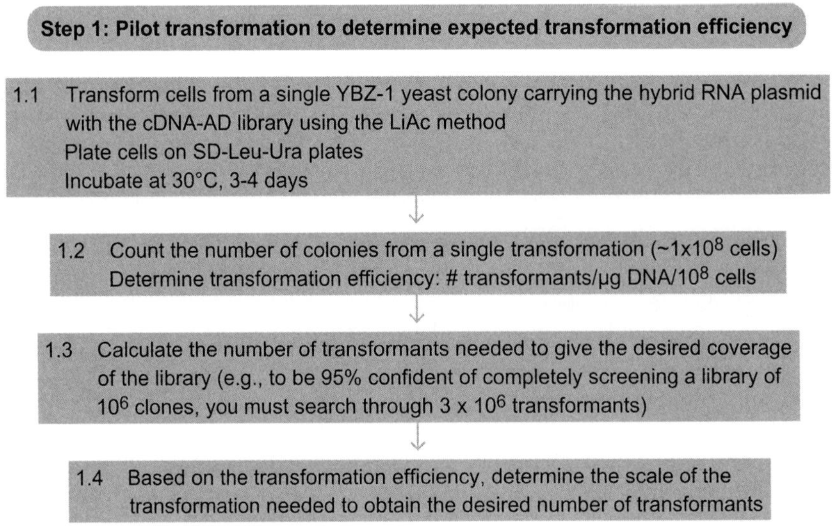

Step 1: Pilot transformation to determine expected transformation efficiency

1.1 Transform cells from a single YBZ-1 yeast colony carrying the hybrid RNA plasmid with the cDNA-AD library using the LiAc method
Plate cells on SD-Leu-Ura plates
Incubate at 30°C, 3-4 days

↓

1.2 Count the number of colonies from a single transformation (~1x10^8 cells)
Determine transformation efficiency: # transformants/μg DNA/10^8 cells

↓

1.3 Calculate the number of transformants needed to give the desired coverage of the library (e.g., to be 95% confident of completely screening a library of 10^6 clones, you must search through 3 x 10^6 transformants)

↓

1.4 Based on the transformation efficiency, determine the scale of the transformation needed to obtain the desired number of transformants

Figure 11.4 Flowchart of Step 1.

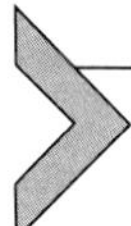

6. STEP 2 DETERMINE 3-AT CONCENTRATION TO BE USED IN SELECTION

6.1. Overview

3-AT is a competitive inhibitor of the *HIS3* gene product and is typically added to select for strong interactions. In this step, one can determine how much 3-AT to include in the selection, such that a balance between suppressing background activation and permitting growth of 'real' positives can be achieved.

6.2. Duration

4–6 days

2.1 Prepare a series of selective medium plates that lack histidine (SD–Leu–Ura–His) and contain increasing amounts of 3-AT (e.g., 0, 0.5, 1, 2, 3, and 5 mM). Also prepare selective medium plates containing histidine (SD–Leu–Ura). The SD–Leu–Ura–His plates are used to assess the level of background activation by the RNA in the absence of a protein. The SD–Leu–Ura plates serve as a control for the transformation.

2.2 Transform cells from a single yeast colony carrying the RNA plasmid with an 'empty' activation domain plasmid. Plate transformants on the prepared plates from Step 2.1. Incubate at 30 °C for 3–5 days.

2.3 Determine the minimal concentration of 3-AT that resulted in no growth of yeast transformants. That would be a suitable starting concentration for the selection experiment.

6.3. Tip

3-AT is light-sensitive. Keep plates in the dark when possible.

6.4. Tip

If colonies are observed even at high concentrations of 3-AT, the RNA likely autoactivates the reporter even in the absence of an interacting protein and is not suitable for the selection experiment.

6.5. Tip

The addition of 3-AT to the medium decreases the number of false positives by demanding a more stringent interaction between the RNA and protein. However,

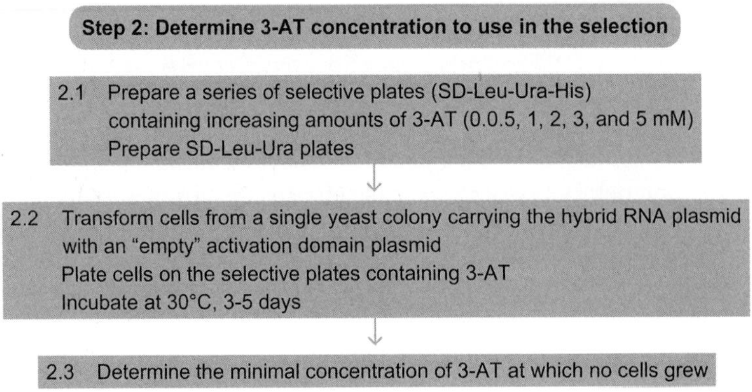

Step 2: Determine 3-AT concentration to use in the selection

2.1 Prepare a series of selective plates (SD-Leu-Ura-His) containing increasing amounts of 3-AT (0.0.5, 1, 2, 3, and 5 mM)
Prepare SD-Leu-Ura plates

↓

2.2 Transform cells from a single yeast colony carrying the hybrid RNA plasmid with an "empty" activation domain plasmid
Plate cells on the selective plates containing 3-AT
Incubate at 30°C, 3-5 days

↓

2.3 Determine the minimal concentration of 3-AT at which no cells grew

Figure 11.5 Flowchart of Step 2.

3-AT concentrations that are too high can result in the loss of colonies containing legitimate RNA–protein interactions.

See Fig. 11.5 for the flowchart of Step 2.

7. STEP 3 INTRODUCE THE cDNA LIBRARY

7.1. Overview

The cDNA library is introduced into YBZ-1 yeast which has been transformed with the hybrid RNA plasmid. Transformants are selected on SD–Leu–His and containing a predetermined amount of 3-AT from Step 2.

7.2. Duration

7–10 days

3.1 Transform cells from a single yeast colony carrying the RNA plasmid with the activation domain library using the LiAc method (Gietz and Woods, 2002, see Chemical Transformation of Yeast). Based on the expected transformation efficiency from Step 1 and the number of transformants desired, the scale of transformation should be adjusted accordingly.

3.2 Calculate the total number of transformants. Save 50 μl of transformed cells from Step 3.1 and make serial dilutions in sterile water (10^{-1}, 10^{-2}, 10^{-3}, and 10^{-4}). Plate 100 μl per dilution on SD–Leu–Ura medium and incubate at 30 °C for 3–4 days. Once colonies are visible, count the number of colonies and calculate the total number of transformants. The confidence level at which the entire library is screened can be calculated using the formula in Step 1.3.

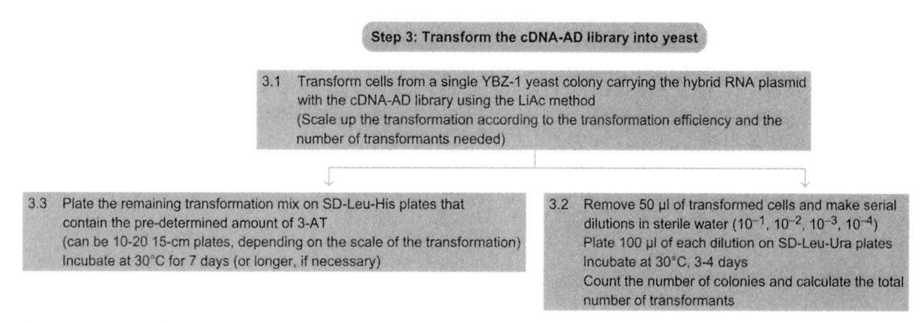

Figure 11.6 Flowchart of Step 3.

3.3 Plate the remaining transformants on SD–Leu–His medium which contains a predetermined amount of 3-AT from Step 2. Incubate at 30 °C for 7 days (or more, if necessary) until colonies start to form.

7.3. Tip

Maintenance of the RNA plasmid (i.e., selection on URA3) is not demanded. This permits cells that can activate HIS3 independent of the RNA to lose the RNA plasmid, permitting the colony color screen below.

See Fig. 11.6 for the flowchart of Step 3.

8. STEP 4 ELIMINATE RNA-INDEPENDENT FALSE POSITIVES BY COLONY COLOR

8.1. Overview

Two classes of positives are obtained from the initial transformation: RNA-dependent and RNA-independent. To facilitate elimination of the RNA-independent false positives, we exploit the *ADE2* gene on the pIIIA hybrid RNA plasmid. The host strain is an *ade2* mutant. When the level of adenine in the medium is low, cells attempt to synthesize adenine *de novo* and accumulate a red purine metabolite due to lack of the *ADE2*-encoded enzyme. This accumulation renders the cells pink or red in color. In contrast, cells carrying the wild-type *ADE2* gene are white.

8.2. Duration

1–2 days

4.1 RNA-independent positives do not require the RNA plasmid to activate the *HIS3* gene and so can lose the plasmid. These false positives will yield pink colonies or white colonies with pink sectors. These colonies should be discarded.

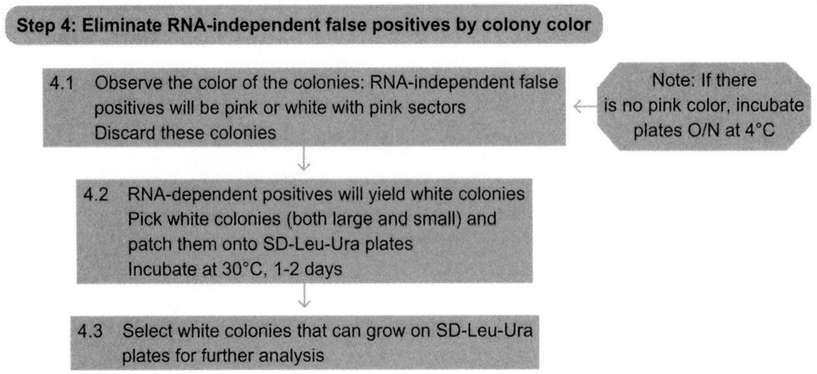

Step 4: Eliminate RNA-independent false positives by colony color

4.1 Observe the color of the colonies: RNA-independent false positives will be pink or white with pink sectors
Discard these colonies

Note: If there is no pink color, incubate plates O/N at 4°C

4.2 RNA-dependent positives will yield white colonies
Pick white colonies (both large and small) and patch them onto SD-Leu-Ura plates
Incubate at 30°C, 1-2 days

4.3 Select white colonies that can grow on SD-Leu-Ura plates for further analysis

Figure 11.7 Flowchart of Step 4.

4.2 For RNA-dependent positives, selection for *HIS3* indirectly selects for maintenance of the RNA plasmid, which carries the *ADE2* gene; thus, these transformants are white. Pick the white colonies (typically a few large and many small colonies) and patch onto SD–Leu–Ura plates to select for the cDNA and RNA plasmids.

4.3 White colonies that are able to grow are selected for further analysis in Step 5.

8.3. Tip

Incubation of the initial transformation plates at 30 °C usually allows enough time for the color to develop. If color development is not strong after a week, incubation of the plates at 4 °C overnight can help.

8.4. Tip

Most of the small white colonies will turn out to be RNA-independent false positives and fail to grow.

See Fig. 11.7 for the flowchart of Step 4.

9. STEP 5 ASSAY β-GALACTOSIDASE ACTIVITY

9.1. Overview

As a secondary selection, the level of β-galactosidase activity is determined by performing a qualitative β-galactosidase activity.

9.2. Duration

1–2 days

5.1 Place a nitrocellulose filter on a plate containing selective medium (SD–Leu–Ura) and patch transformants directly on the filter. Incubate cells at 30 °C for 1–2 days.

5.2 Lyse cells by immersing the filter in liquid nitrogen for 5–20 s and thawing on the benchtop for 1 min.

5.3 Repeat freeze/thaw cycle twice.

5.4 Place a piece of Whatman 3-MM chromatography paper in a Petri dish and saturate it with 5-ml Z-buffer. Add 75 μl of 20 mg ml^{-1} X-Gal solution. Discard excess buffer in Petri dish.

5.5 Overlay the nitrocellulose filter containing cells onto the saturated Whatman paper, cells side up. Seal dish with parafilm.

5.6 Incubate, colony side up, 30 min to overnight at 30 °C. Examine the filters regularly. A strong interaction (such as that between IRE and IRP) should turn blue within 30 °min.

9.3. Tip

Typically, 50 positives can be patched onto a 100-mm diameter Petri dish.

9.4. Caution

After freezing and thawing, the nitrocellulose filter is brittle and may crack easily.

9.5. Tip

X-Gal should be added fresh. Keep stock solution in the dark as it is light-sensitive.

9.6. Tip

Ensure good contact between the nitrocellulose filter and the Whatman paper.

9.7. Tip

With protracted incubation, weak interactions eventually yield a blue color. For this reason, it is important to examine the filters periodically to determine how long it takes for the color to develop. It is also helpful to include a positive control (e.g., IRE–IRP interaction) to ensure that the assay works.

See Fig. 11.8 for the flowchart of Step 5.

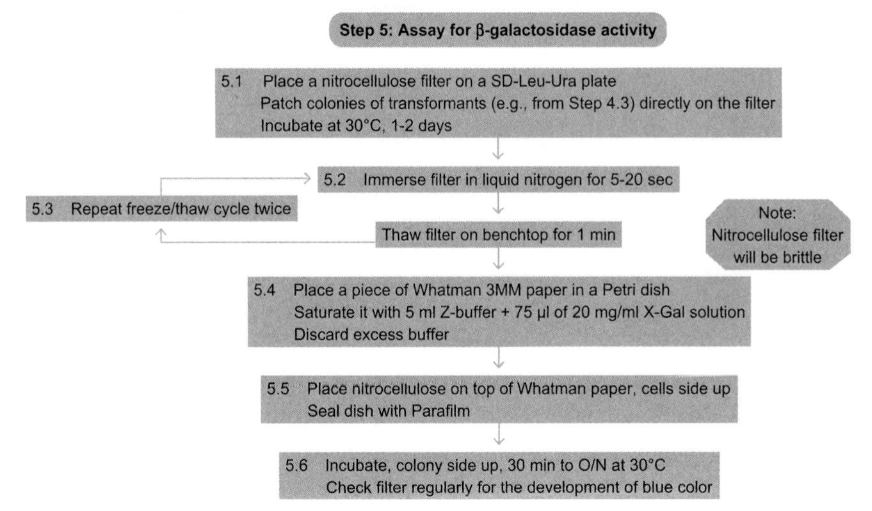

Figure 11.8 Flowchart of Step 5.

10. STEP 6 CURE THE RNA PLASMID AND TEST POSITIVES FOR RNA DEPENDENCE

10.1. Overview

Most, but not all, of the RNA-independent false positives are eliminated by the colony color assay in Step 2. To ensure that the positives are RNA-dependent, the RNA plasmid is removed by counterselection against *URA3*. Cells will be plated on media containing 5-fluoroorotic acid (5-FOA). 5-FOA is converted by the *URA3* gene product to 5-fluorouracil, which is toxic. Cells lacking *URA3* can grow in the presence of 5-FOA if uracil is provided, while cells containing *URA3* cannot. Expression of *lacZ* is then monitored. Candidates that fail to activate the reporter genes are further analyzed.

10.2. Duration

4–5 days

6.1 Replica-plate the positives from Step 4 to SD–Leu plates. Incubate the plates at 30 °C for 1 day, allowing the cells to lose the RNA plasmid.

6.2 Replica-plate the cells from Step 6.1 onto SD–Leu + 0.1% 5-FOA plates. Incubate the plates at 30 °C for 2–3 days.

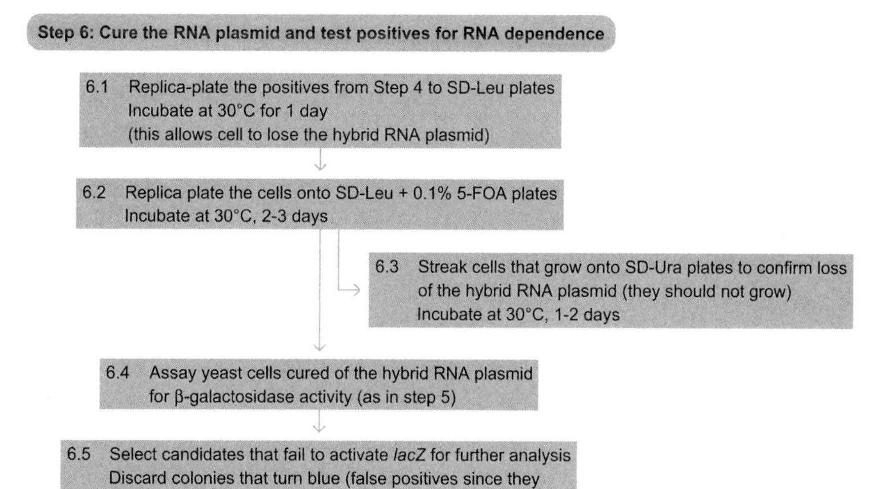

Figure 11.9 Flowchart of Step 6.

6.3 Cells that grow can be streaked on SD–Ura plates to confirm the loss of the RNA plasmid. A single pass through 5-FOA counterselection is usually sufficient.

6.4 Assay β-galactosidase activity as in Step 5.

6.5 Select the candidates that fail to activate *lacZ* for further analysis. Discard false positives which activate *lacZ* (blue color) in the absence of the RNA plasmid.

See Fig. 11.9 for the flowchart of Step 6.

11. STEP 7 ISOLATE PLASMIDS FOR AUTOACTIVATION TEST AND SEQUENCING

11.1. Overview

Once candidates are identified, the cDNA plasmid must be recovered from yeast and introduced into *E. coli* (see Transformation of Chemically Competent *E. coli* or Transformation of *E. coli* via electroporation). The plasmids can then be amplified in *E. coli* for use in future applications. cDNA plasmids must be transformed back into a yeast strain containing an 'empty' RNA plasmid to test for RNA-independent activation (see Chemical Transformation of Yeast). Expression of either the *HIS3* or *lacZ* reporter gene is monitored.

11.2. Duration

3–5 days

7.1 Inoculate each positive clone (from patched positives in Step 6.2) in 5-ml SD–Leu medium and grow to saturation (24–48 h) at 30 °C in a rotary shaker (200 rpm).

7.2 Pellet 2-ml cells at 5000 × *g*. Resuspend the pellet in 250 μl of Cell Resuspension Buffer.

7.3 Spheroplast yeast with 5-μl Zymolyase for 2 h at 37 °C.

7.4 Add 250-μl lysis buffer and invert to mix.

7.5 Incubate for 5 min at room temperature, followed by 5 min at 65 °C. Cool to room temperature.

7.6 Add 10 μl of alkaline protease solution and incubate for 10 min at room temperature.

7.7 Add 350 μl of lysis buffer and invert to mix.

7.8 Centrifuge at >16 000 × *g* for 10 min.

7.9 Add lysate to the binding column. Centrifuge at >16 000 × *g* for 1 min.

7.10 Wash the column with 700-μl wash buffer. Centrifuge at 16 000 × *g* for 1 min. Discard the flow-through.

7.11 Wash the column with 500-μl wash buffer. Centrifuge at >16 000 × *g* for 1 min. Discard the flow-through.

7.12 Centrifuge at >16 000 × *g* for 2 min to remove the residual buffer.

7.13 Elute the plasmid DNA in 100-μl water.

7.14 Transform the plasmid DNA into *E. coli* using conventional methods. Plate transformants on medium containing the appropriate antibiotic to select for the cDNA plasmid.

7.15 Perform colony PCR to identify clones carrying the cDNA plasmid (see Colony PCR). Isolate plasmids from *E. coli* using conventional methods.

7.16 Transform each cDNA plasmid back into a yeast strain containing an 'empty' RNA plasmid.

7.17 Assay transformants for the expression of the *lacZ* reporter, as described in Step 5.

7.18 Discard false positives that activate *lacZ* in the absence of your specific RNA sequence. Identify the remaining cDNAs by sequencing the plasmids isolated from *E. coli*.

11.3. Tip

This step is a modified version of the Promega Wizard Plus SV Miniprep kit.

11.4. Tip

Since either RNA or cDNA plasmids can be present in E. coli, *colonies can be screened for the plasmid of interest by colony PCR. If different antibiotic resistance markers are present on each plasmid, most colonies should contain the cDNA plasmid.*

11.5. Tip

Alternatively, HIS3 expression can be monitored as described in Dissecting a known RNA-protein interaction using a yeast three-hybrid system.

11.6. Tip

Compare the cDNA sequences to common sequence databases to identify protein partners (e.g., ***BLAST****).*

See Fig. 11.10 for the flowchart of Step 7.

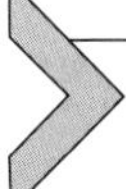

12. STEP 8 DETERMINE BINDING SPECIFICITY USING MUTANT AND CONTROL RNAs

12.1. Overview

Clones that require the RNA plasmid to activate the reporter genes can then be analyzed to determine whether they interact sequence specifically with the RNA.

12.2. Duration

3–4 days

8.1 Introduce small, directed (point) mutations in the hybrid RNA (e.g., mutating a nucleotide of known importance in the RNA) that disrupt the interaction between the RNA and the protein (see Site-Directed Mutagenesis).

8.2 Transform the candidate cDNA plasmids into a yeast strain containing either the wild-type or mutant hybrid RNA plasmids.

8.3 Assay transformants for the expression of the *lacZ* reporter qualitatively, as described in Step 5, to determine the sequence specificity of the positive clones.

12.3. Tip

In this case, it may be helpful to know the identity of the candidates from the screen and use this information to guide mutagenesis studies.

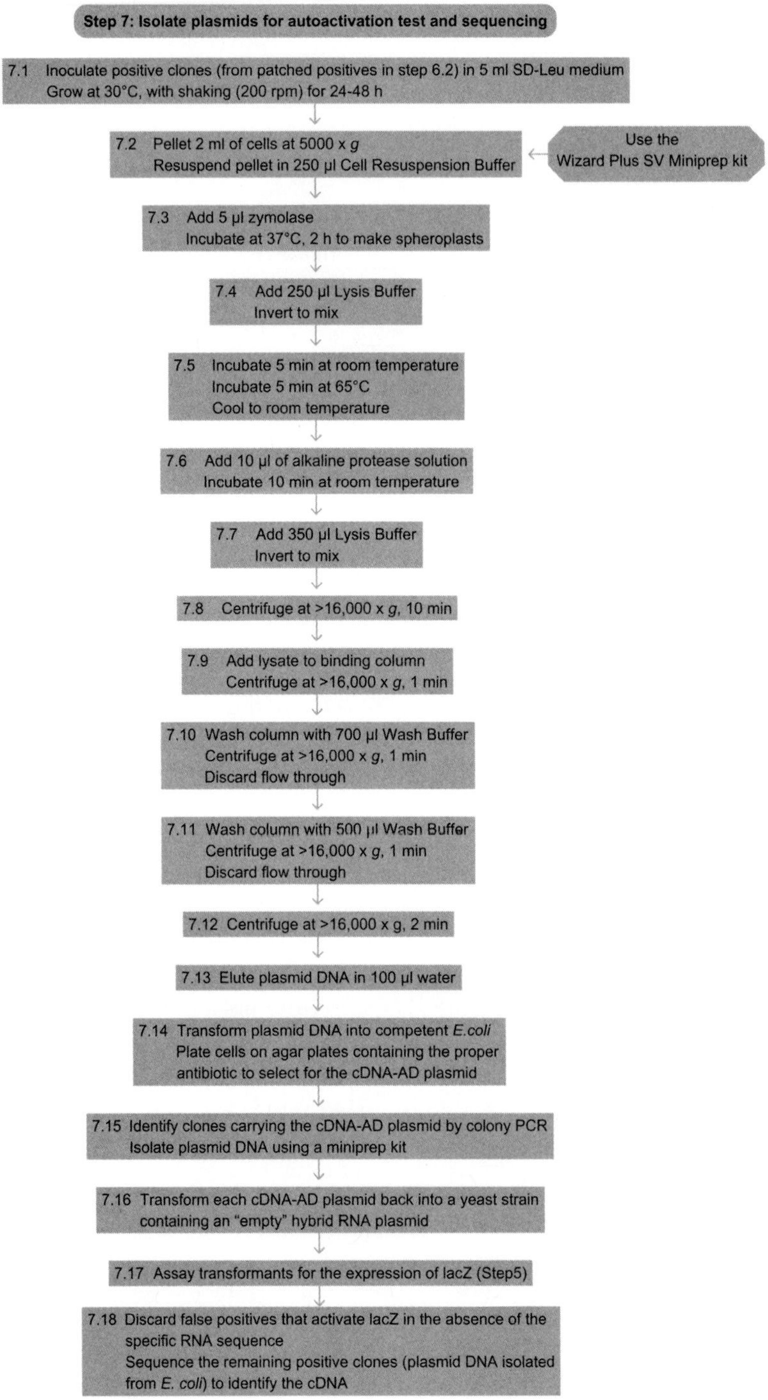

Step 7: Isolate plasmids for autoactivation test and sequencing

7.1 Inoculate positive clones (from patched positives in step 6.2) in 5 ml SD-Leu medium
Grow at 30°C, with shaking (200 rpm) for 24-48 h

7.2 Pellet 2 ml of cells at 5000 x *g*
Resuspend pellet in 250 µl Cell Resuspension Buffer

Use the Wizard Plus SV Miniprep kit

7.3 Add 5 µl zymolase
Incubate at 37°C, 2 h to make spheroplasts

7.4 Add 250 µl Lysis Buffer
Invert to mix

7.5 Incubate 5 min at room temperature
Incubate 5 min at 65°C
Cool to room temperature

7.6 Add 10 µl of alkaline protease solution
Incubate 10 min at room temperature

7.7 Add 350 µl Lysis Buffer
Invert to mix

7.8 Centrifuge at >16,000 x *g*, 10 min

7.9 Add lysate to binding column
Centrifuge at >16,000 x *g*, 1 min

7.10 Wash column with 700 µl Wash Buffer
Centrifuge at >16,000 x *g*, 1 min
Discard flow through

7.11 Wash column with 500 µl Wash Buffer
Centrifuge at >16,000 x *g*, 1 min
Discard flow through

7.12 Centrifuge at >16,000 x g, 2 min

7.13 Elute plasmid DNA in 100 µl water

7.14 Transform plasmid DNA into competent *E.coli*
Plate cells on agar plates containing the proper antibiotic to select for the cDNA-AD plasmid

7.15 Identify clones carrying the cDNA-AD plasmid by colony PCR
Isolate plasmid DNA using a miniprep kit

7.16 Transform each cDNA-AD plasmid back into a yeast strain containing an "empty" hybrid RNA plasmid

7.17 Assay transformants for the expression of lacZ (Step5)

7.18 Discard false positives that activate lacZ in the absence of the specific RNA sequence
Sequence the remaining positive clones (plasmid DNA isolated from *E. coli*) to identify the cDNA

Figure 11.10 Flowchart of Step 7.

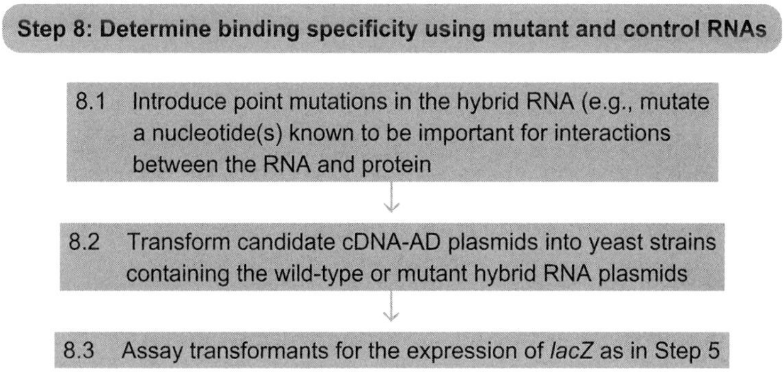

Step 8: Determine binding specificity using mutant and control RNAs

8.1 Introduce point mutations in the hybrid RNA (e.g., mutate a nucleotide(s) known to be important for interactions between the RNA and protein

↓

8.2 Transform candidate cDNA-AD plasmids into yeast strains containing the wild-type or mutant hybrid RNA plasmids

↓

8.3 Assay transformants for the expression of *lacZ* as in Step 5

Figure 11.11 Flowchart of Step 8.

12.4. Tip

If no subtle mutations are available, rudimentary analyses such as using deletions or antisense RNA can be helpful.

12.5. Tip

If a quantitative comparison between protein and various RNAs is desired, a quantitative β-galactosidase assay is provided in Dissecting a known RNA-protein interaction using a yeast three-hybrid system.

See Fig. 11.11 for the flowchart of Step 8.

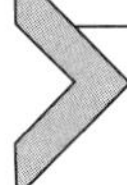

13. STEP 9 FUNCTIONAL TESTS OR ADDITIONAL SCREENS

13.1. Overview

Almost invariably, additional steps will be needed to identify those positives that are biologically meaningful. Each screen is unique. The interactions being analyzed and the organisms being studied will determine what additional steps need to be taken to determine the biological relevance of each interaction. It is not surprising, given that the assay is performed outside of most biological contexts, that some specific, high-affinity interactions may not be relevant to the biology of the system being studied.

REFERENCES

Referenced Literature

Gietz, R. D., & Woods, R. A. (2002). Transformation of yeast by lithium acetate/single-stranded carrier DNA/polyethylene glycol method. *Methods in Enzymology*, *350*, 87–96.

Stumpf, C. R., Opperman, L., & Wickens, M. (2008). Chapter 14. *Analysis of RNA-protein interactions using a yeast three-hybrid system Methods in Enzymology*, *449*, 295–315.
Zhang, B., Gallegos, M., Puoti, A., et al. (1997). A conserved RNA-binding protein that regulates sexual fates in the C. elegans hermaphrodite germ line. *Nature*, *390*, 477–484.

SOURCE REFERENCES

Bernstein, D. S., Buter, N., Stumpf, C., & Wickens, M. (2002). Analyzing mRNA-protein complexes using a yeast three-hybrid system. *Methods*, *26*, 123–141.
Stumpf, C. R., Opperman, L., & Wickens, M. (2008). Chapter 14. *Analysis of RNA-protein interactions using a yeast three-hybrid system Methods in Enzymology*, *449*, 295–315.

Referenced Protocols in Methods Navigator

Chemical Transformation of Yeast.
Saccharomyces cerevisiae Growth Media.
Transformation of Chemically Competent *E. coli*.
Transformation of *E. coli* via electroporation.
Colony PCR.
Dissecting a known RNA-protein interaction using a yeast three-hybrid system.
Site-Directed Mutagenesis.

AUTHOR INDEX

Note: Page numbers followed by "*f*" indicate figures.

K

L

M

N

O

P

R

S

T

U

V

W

Y

Z

SUBJECT INDEX

Note: Page numbers followed by "*f*" indicate figures.